CRIME, POLICING AND PLACE

Spiralling crime rates and continuing public concern about police–community relations ensure that crime and policing remain firmly on the social and political agenda. An awareness of crime continues to affect the lives of ordinary people and also to stimulate policy-makers who recognize that crime rates form one of the litmus papers by which their effectiveness is judged. Of the many agencies involved in the battle against crime, the police in their various roles constitute the most obvious front line.

Drawing in case material from Britain, Europe, Canada and America, *Crime, Policing and Place* examines the significance of spatial patterns of crime and the processes which produce them. The book analyses the implications of theoretical and methodological innovation in the study of crime and policing, the processes which underlie the uneven distribution and impact of crime and the success of recent policies aimed at preventing crime and enhancing police–community relations.

The contributors are drawn from a variety of academic disciplines which include criminology, geography and social policy and also from police and government agencies with direct inputs into policy.

David Evans is Senior Lecturer in Geography at Staffordshire Polytechnic. **Nicholas Fyfe** is Lecturer in Geography at Strathclyde University. **David Herbert** is Professor of Geography at the University College of Swansea.

CRIME, POLICING AND PLACE

Essays in environmental criminology

Edited by
David J. Evans, Nicholas R. Fyfe
and David T. Herbert

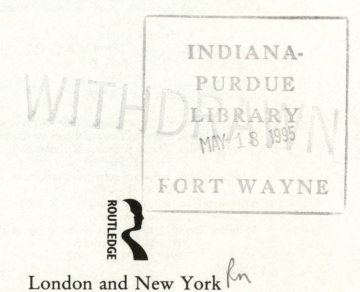

ROUTLEDGE

London and New York

First published 1992
by Routledge
11 New Fetter Lane, London EC4P 4EE

Simultaneously published in the USA and Canada
by Routledge
a division of Routledge, Chapman and Hall, Inc.
29 West 35th Street, New York, NY 10001

Typeset by J&L Composition Ltd, Filey, North Yorkshire
Printed and bound in Great Britain by
Mackays of Chatham PLC, Chatham, Kent

British Library Cataloguing in Publication Data
Crime, policing and place: essays in environmental
criminology
I. Herbert, David, *1935–*
II. Evans, David J. (David John), *1947–*
III. Fyfe, N. R.
364.49
ISBN 0–415–04990–3

Library of Congress Cataloging in Publication Data
Crime, policing, and place: essays in environmental
criminology/edited by David J. Evans, Nicholas R. Fyfe, and
David T. Herbert.
Includes bibliographical references and index.
ISBN 0–415–04990–3
1. Criminology.   2. Human geography.   3. Crime – Case
studies.   4. Crime prevention – Case studies.
I. Evans, David J., 1947–   II. Herbert, David T.
III. Fyfe, Nicholas R.
HV6030.C742   1992
364.2′2—dc20   91–24980
                                                  CIP

FTW
AGE 0950

CONTENTS

v

CONTENTS

FIGURES

FIGURES

TABLES

TABLES

CONTRIBUTORS

Robert Barr	Department of Geography, University of Manchester
Trevor Bennett	Institute of Criminology, University of Cambridge
Anthony E. Bottoms	Institute of Criminology, University of Cambridge
Ann Claytor	Centre for Criminological and Legal Research, University of Sheffield
Judy Darwood	Department of Geography, University of Wales (Swansea)
Norman Davidson	School of Geography and Earth Resources, University of Hull
David J. Evans	Department of Geography and Recreation Studies, Staffordshire Polytechnic
Nicholas R. Fyfe	Department of Geography, University of Strathclyde
Neil Hall	Northumbria Police Force, Ponteland
Kevin Heal	Home Office Crime Prevention Unit, London
David T. Herbert	Department of Geography, University of Swansea
Ian Heywood	Department of Geography, Salford University
János Ladányi	Department of Sociology, University of Economics, Budapest, Hungary
Michael Levi	School of Social and Administrative Studies, University of Wales (Cardiff)
Trevor Locke	Policy Planning Officer, Coventry City Council

CONTRIBUTORS

John Lowman   Department of Criminology, Simon Fraser
                       University, Canada
David Mitchell   Department of Geography, Edinburgh
                       University
Kate Painter   Department of Social Policy, University of
                       Manchester
Ken Pease   Correctional Service of Canada, University
                       of Saskatchewan
Peter Redhead   Gateshead Multi-Agency Crime Prevention
George F. Rengert   Department of Criminal Justice, Temple
                       University, Philadelphia, USA
Paul Wiles   Faculty of Law, University of Sheffield

INTRODUCTION

David J. Evans, Nicholas R. Fyfe and David T. Herbert

As the interface between crime and policing is of critical importance in the fight against rising crime rates, so analysis intended to provide a clearer understanding of this interface must rank highly in the research priorities of criminological studies. Academics have frequently raised serious questions on the roles of the police. The efficacy of the data which the police collect and which form the bases for official crime statistics has been and remains a major issue in its own right. Similarly the effectiveness of policing has been questioned by criminologists; evaluations of specific policing strategies remain ambiguous in their main findings. Much of this criticism has proved difficult to counter but it should also be recognized that academics have been distinctly coy in terms of bringing forward alternative proposals. It should also be recognized that at various points in the methodological debates on the sources of crime, the 'us and them' dichotomy has placed the official agencies of law and order on the 'them' side of the debate. There are now many more signs of constructive collaboration and the chapters in this book will reflect these changes. Academics can show that there are more efficient ways of collecting crime data which, while they will not eliminate all of the problems, are steps in the right direction. Some of the major initiatives in crime prevention can be linked to findings of academic research which when put into practice provide positive results. Both these examples provide connections with the other special feature of this collection of essays: they are concerned with new spatial perspectives and form part of a specialism which can be labelled as environmental criminology. The more efficient ways of collecting data are those emanating from the impact of geographic information systems (GIS) which is a more general set of procedures but one well suited to the collecting

1

and storing of criminal statistics. One of the most widely applied and tested crime prevention projects is that known variously as neighbourhood watch, community watch or block watch. It involves an area policy in which residents are encouraged to act as the 'eyes and ears' of their community in the fight against crime. As a general concept, the area policy has roots in social geography and its concern with place.

It has been obvious for a very long time that there are geographies of crime in the sense that offenders and offences show marked tendencies to concentrate in particular sections of geographical space and these features are well documented (see Herbert 1982). It has also been shown that as there is no uniform distribution in space of crime so there is no uniformity in the spatial impacts of laws and justice (see Harries and Brunn 1978). Both of these observations now have the status of truisms and the research task is not that of proving the existence of spatial patterns but rather of examining their significance and of understanding the processes which produce them. In many ways the geography of crime has moved away from a focus on space and pattern, from what is sometimes termed a preoccupation with the 'fetishism of space', and in this way reflects the changing concerns of the wider discipline of which it is part. This in its way creates new tensions. The identification of 'areas at risk', 'hot-spots', 'vulnerable areas' and 'problem estates' may become mechanical and rather basic in its concept and method, but the reality of the crime problem remains acute for people who occupy these parts of the city. If there is a tension it is between the search for deeper understanding and the imperative of short-term amelioration; the real research task is to bring these into harmony.

The contributions in this book are seen as steps in these directions. They advocate the value of a dimension within criminology which is not negligent of the roles of place and the spatial perspective; they portray this dimension in ways very different from the older ecological perspective. Environmental criminology as exemplified by these contributions has both an academic rigour and a practical application and these two objectives need not be exclusive. The book is also importantly not solely a collection of essays by geographers. Some of the most notable historical contributions to the *geography* of crime have come from outside the discipline of geography; the origins of cartographic criminology in the work of statisticians and of the Chicago school of social ecology stand as notable exemplars. In recent years the Sheffield study emanating from a centre for criminological and legal research has become the major area-based analysis of criminality which continues to make significant contributions to this field of study. Environmental criminology and the wider concept of the geography of crime are not held in monopoly by geographers and this evidence of shared concerns and activities, mirrored in this book, is a welcome and positive sign of progress.

For a book which is intended to demonstrate the changing interests of geographers in the study of crime and policing, it was felt appropriate that two non-geographers should have the initial word and a brief to evaluate the researches into crime and place. Anthony Bottoms and Paul Wiles have both played major roles in the Sheffield studies of crime and much of their recent writing has involved contact with environmental criminology. Both are leading criminologists in their own right and have wider interests than crime and place; they are in many ways ideally placed to offer a state-of-the-art review. The first chapter in this book under their authorship has the title 'Explanations of crime and place' and bids fair to be a definitive and substantive statement of great value for all with research interests in this area. Bottoms and Wiles consider crime and place through the theory of structuration and the ways in which this provides insights into the intimate relationships between social relations and spatial structures. Structural theory has more general application within social geography but has hitherto received little attention in analyses of the spatial dimensions of crime and offending. Within the whole field of criminology there is an inconsistent awareness of the relevance of a spatial dimension and this neglect is evident in recent publications. In part at least this neglect can be attributed to the fact that the spatial dimension is inadequately theorized. This problem is implicit even in major projects such as those of the Chicago school and becomes most obvious on attempting to bridge the gap between spatial patterns and ethnographic studies. Space and time are central to the Giddens theory of structuration and this suggests that the approach has a great deal to offer environmental criminology. Bottoms and Wiles demonstrate the value of structuration theory by introducing it to a recent study of urban crime in Sweden. They conclude that the Swedish approach markedly understates the importance of social processes and, by paying too little attention to the routine activities, perceptions and decisions of individual actors, risks missing the distinction, central to structuration theory, between the intended

3

and unintended consequences of individual action. Reiss argued that neighbourhood crime patterns could be thought of as analogous to communities having crime careers. This concept of a community crime career embodies the idea of social change at a meso-scale. The implications from the Giddens structuration theory are that exact prediction of the course of such careers is impossible but factors influencing their development can be analysed in probabilistic terms (see Gregory and Urry 1985). At a time when many significant place-changes are evident in the British urban system, their criminological consequences raise a critical research agenda for environmental criminologists. Continuing the first part of the book, which focuses on methodologies and sources, David Evans reviews the left realist position, which has attracted attention in recent years. The approach is to select a number of themes – data sources, victimization experience, fear and causes of crime and crime prevention policies – and to consider the ways in which the left realism perspective throws new light upon them. As a generalization the chapter concludes that left realism provides a body of theory which can assist in the spatial analysis of crime.

The final two chapters in the first part of the book are concerned with the handling of crime data. Norman Davidson and Trevor Locke advocate the need for local crime profiles and demonstrate how these may be developed and used. Using a 'difference-model' the authors compare actual local crime levels with those which could be anticipated from national and regional incidence rates. The difference model is a variant of the well-used shift-share analysis and is an alternative to more traditional location quotients and similarity indices. Application of the model details the crime profile of the chosen area but an essential adjunct is a parallel community profile. Profiling can be used to draw conclusions about the impact of policing and crime prevention measures and is crucial to community-based anti-crime initiatives. Ian Heywood, Neil Hall and Peter Redhead offer a more general example of a geographic information systems (GIS) application to crime data. Arising from the Five Towns Initiative and the multi-agency approach in North Tyneside, the crime file data were referenced at the scales of estate, street and postcode. Data inputs came from residential and commercial surveys but only selectively from less well referenced police data. The files were used to target crime prevention policies and had positive effects. Some of the technical difficulties are reviewed and the problems and prospects for developing GIS applications in this area are evaluated.

The second part of the book contains eight of the sixteen contributions and is rather more diverse in its content. János Ladányi has written a more traditional form of ecological analysis of offenders in Budapest, using a data set of imprisoned offenders. Basically he examines a number of distributions by aggregate districts within the city and searches for meaningful correlations. Methods are conventional and findings are tentative but the study does offer an insight into crime patterns in a part of eastern Europe which is thinly documented in these terms. George Rengert has written extensively on the journey to crime and in his chapter draws together some of the key findings in the literature. The focus in this chapter is upon public policy issues and their impact upon the distribution of criminality. At an aggregate level, any public policy which increases the level of criminality such as early release programmes contributes to spatial injustice in that those who suffer most from victimization are once again the likely victims. At an individual level, drug sale locations are becoming nodal points for the journey to crime; any policy which 'enables' drug sales at particular locations affects the geography of crime. Again, housing policies in the public sector and levels of security have an impact upon the journey to crime and official responsibility for displacement of crime needs to be monitored. Anthony Bottoms, Ann Claytor and Paul Wiles offer a new empirical outcome of research on the Sheffield crime study. Using the concept of residential community crime career first developed by Albert Reiss, they apply this to two housing estates in Sheffield; their central claim is that in order to understand and explain offending behaviour by residents of particular areas, it is vital to consider who lives there, how they got there, what kinds of social life they lead, how outside agencies regard them and why they stay. The Sheffield database contained a wealth of information not only on crime and social demography but also of great significance is the housing allocation system, which can crucially affect the development of community crime careers. Residential area or neighbourhood remains the theme for a study of crime awareness by David Herbert and Judy Darwood. With the advent of national crime surveys, the significance of fear of crime and its effects on people's lives has become more widely recognized and this study of Swansea neighbourhoods explores some of these relationships at a locality scale. Tests on perception of crime levels showed an accurate assessment relative to recorded crime rates for small areas. Awareness of crime affected satisfaction with living in particular neighbourhoods and also security consciousness. Crime prevention programmes heighten awareness of crime and run the risk of heightening fear, particularly among those groups such as elderly people and women who feel they are the most vulnerable.

This idea of vulnerable groups is developed in detail in Kate Painter's analysis of female victimization and her characterization of the two worlds in which men and women live. There is a growing volume of evidence from both national and local surveys which throws the vulnerability, perceived and real, of women to crime in sharper focus. There are ways of reducing risk but many of these involve curtailment of freedom of action and activities and should not be over-regarded as acceptable options. Much of the extant situation and its posited solutions emanates from the nature of a society which is male dominated and is defined, consciously or subconsciously, in those terms. Robert Barr and Ken Pease develop some of their new ideas in relation to displacement of crime. Reviewing different types of displacement they prefer the term 'deflection' which involves moving crime from a chosen target. The analysis of 'crime flux' focuses on crime movements across offenders and crime types and its consequences may be benign, malign or neutral. In order to demonstrate the use of the crime-flux approach, a notional classification using indices of concentration and prevalence is used: British Crime Survey data show a strong relationship between these two measures. From the classification a range of types of 'shifts' can be discerned and interpreted with reference to specific crimes. The conceptual framework of crime flux stands in need of further research and refinement but offers significant advantages over traditional displacement models.

White-collar crime has been an issue in criminology for half a century but remains relatively unstudied. As the great majority of criminology continues to concentrate on crimes of the common people, the criticism that it ignores the crimes of the wealthy and powerful continues to be heard. Michael Levi has a fine record of research into white-collar crime in its various forms and his chapter throws new light and insight on to a neglected and difficult area. One focus is the policing of fraud and the problems which that poses in cross-cultural circumstances. Judicial systems are often ill-equipped to deal with complex situations and new organizations are only slowly evolving. International co-operation, inter-regional agencies and 'national' police forces are all mooted *desiderata* in the fight against white-collar crime. While there are moves in these directions, such as the internationalization of fraud regulation, the motives are not always unambiguous and 'enlightenment prospers best when allied with self-interest'. Whereas Michael Levi is concerned with the 'global village', John Lowman's discussion of policing practice returns us to the local scale. Lowman has the view that police practices have limited importance in understanding intra-urban crime rates but do influence the location of offences and the kinds of people who are processed as criminals, especially for victimless crime. Drawing upon the example of street prostitution in Vancouver, it can be shown that the changing geography has resulted directly from strategies of law enforcement. Ways in which police have been deployed have led to particular groups being processed by the criminal justice system. Police behaviour and attitudes may well influence patterns of crime and offender residence because policing can be both discretionary and differential.

At a time when crime and policing policies are high on the political agenda and have become an electoral issue, it was appropriate for the original conference at which many of these papers were presented to be addressed by a government minister with responsibility for crime prevention. John Patten, who is also incidentally a former university lecturer in geography, took a keen interest in the issues under debate. His interests are perhaps most closely reflected in the final part of the book, which turns to policy perspectives and includes studies of crime prevention and policing. Kevin Heal, who is head of the Home Office's Crime Prevention Unit, reviews changing British crime prevention initiatives over time and identifies significant shifts in emphases. It was during the 1980s that area profiles were most strongly developed with initiatives such as the Five Towns and Safer Cities projects. Crime prevention is making ground but is still hampered by the absence of good local structures and basic crime pattern analysis. Of the current crime prevention initiatives, neighbourhood watch is the most widely recognized and applied. Trevor Bennett examines the present status of neighbourhood watch schemes in North America and the United Kingdom. Undoubtedly both the take-up and effectiveness of schemes varies with social and demographic characteristics of communities and the charge is often made that they are least evident where they are most needed. The first wave of evaluations were all positive and pointed to crime reductions but there are questions of methodological weaknesses and the reliability of tests. Initial reductions in crime may not be maintained and schemes need to be reconsidered as vehicles for crime prevention.

David Mitchell in his study of police initiatives in Brixton (London) since the 1981 riots and the subsequent Scarman Report, reports on detailed studies with a range of individuals, groups and agencies resident in the area. The Community/Police Consultative Group for Lambeth (CPCGL) was set up in 1982 to foster better relations between police and community; the Lay Visitors scheme gave some members of the public access to police stations; and the community involvement offices again have the task of increasing police involvement in the community. Whereas these are formal agencies with varying levels of success, informal contacts engendering mutual self-respect are at least as important as the work of any committees. This theme is developed more generally with the chapter by Nicholas Fyfe on police/community consultation. Using information from two London boroughs, the chapter explores the effectiveness of consultative committees and questions their real power. There are critical issues surrounding the composition of the committees and their representativeness. The role of senior police officers has implications for management and resource allocation and also raises questions of local autonomy. Despite the similar asymmetrical power relations between the police on the one hand and the community and local elected members on the other, there are strong local variations in the outcomes of community/police consultations.

The sixteen chapters which comprise this book are held together by their concern with crime and policing. They have been arranged within the text to provide a clear structure and a flow of ideas from initially concepts and methods to, ultimately, policies and their implementation. Within this set a striking feature is the emergence of new topics being taken up by geographers and others concerned with environmental criminology. There is a vitality about the research interests which testifies to the continuing momentum of these *new* spatial perspectives as they are applied to studies of criminality and policing practice.

REFERENCES

Gregory, D. and Urry, J. (eds) (1985) *Social Relations and Spatial Structures*, London: Macmillan.

Harries, K.D. and Brunn, S.D. (1978) *The Geography of Laws and Justice*, New York: Praeger.

Herbert, D.T. (1982) *The Geography of Urban Crime*, London: Longman.

Part I

METHODOLOGIES AND SOURCES

1

EXPLANATIONS OF CRIME AND PLACE

Anthony E. Bottoms and Paul Wiles

The opportunity for two criminologists to reflect, in the company of geographers, upon some aspects of the spatial distribution of crime produces something of a dilemma. On the one hand we want to give full weight to and to welcome the very real contributions which geographers have recently made to this subject (e.g. in Britain alone, Davidson 1981; Herbert 1982; Smith 1986); on the other hand, there seems little point in producing yet another substantive overview of what these writers, and their criminological and sociological colleagues, have discovered to date.

We have chosen, therefore, to write largely in a methodological vein, though with reference where appropriate to substantive findings. Our framework of approach is the intimate relationship of social relations and spatial structures found within the theory of 'structuration', a framework which has attracted considerable recent attention and critical debate among general human geographers and social theorists (e.g. Gregory and Urry 1985), but which has so far been given little consideration in discussions of the spatial dimensions of crime and offending. We adopt this approach because structuration theory offers a model for explanation which brings together, in a coherent fashion, a number of elements which we have been developing in our own work on residential areas and crime; it offers, in our view, both a framework within which previous research can be synthesized, and a valuable stimulus for future research.

A useful starting-point is to consider some recent contributions to the criminological literature. The years 1989 and 1990 saw the publication of two ambitious books, one Australian, one American, claiming to offer 'general theories of crime', that is to say, theories that would explain all or most crime in (at the minimum) western

11

societies (Braithwaite 1989; Gottfredson and Hirschi 1990).[1] This is not an appropriate place to discuss the general merits and demerits of these theories; suffice to say that anyone well versed in the literature of 'environmental criminology'[2] would find a remarkable absence of the spatial dimensions of crime and offending in both books. Braithwaite's 'theory of reintegrative shaming' devotes very little attention to *crimes* as opposed to *offenders*,[3] and even offender rate variations are discussed in a largely aspatial manner. Gottfredson and Hirschi's theory takes crimes more seriously, but largely from the perspectives of classical theory and simple opportunity theory, and with little explicit discussion of spatial issues (but see pp. 12–13);[4] in discussing offenders, these authors' key concept is low social control, and again spatial processes are largely absent from the discussion.

By way of contrast consider an article published in the American journal *Criminology* in 1989, on the 'hot spots of predatory crime' (Sherman *et al.* 1989). The authors used 'police call data'[5] for Minneapolis for 1985–6, and discovered among other things

1 that just 3.3 per cent of addresses and intersections in the city generated 50 per cent of all calls to the police for which cars were dispatched
2 there was great variation in the victimization of specific micro-locations even within high crime areas.

From these data Sherman *et al.* pose the question (not answerable with the kind of information they had access to):

> whether the routine activities of places, given their physical environment, are actually criminogenic. Do places vary in their capacity to help *cause* crime, or merely in their frequency of *hosting* crime that was going to occur some place inevitably, regardless of the specific place? Are the routine activities of hot spots criminogenic *generators* of crime, or merely more attractive *receptors* of crime? (Sherman *et al.* 1989: 46, italics in original)

The Minneapolis data certainly seem to suggest that a criminological theory without an adequate dimension of place (such as those of Braithwaite or Gottfredson and Hirschi) will be incomplete – unless, of course, one is prepared to argue that the spatial dimension is purely epiphenomenal, which is possible, but seems unlikely (see pp. 17–19). But the questions posed by the authors about their results in turn create some anxieties. In speaking about 'places [which] *cause* crime', are Sherman and his colleagues adopting an over-deterministic approach to the idea of place?[6] Anxiety is to some extent allayed by the introduction of the concept of 'routine activities' (from Cohen and Felson 1979), which clearly introduces an element of social activities, linked in this case to specific locations. But is the concept of 'routine activities of hot spots' itself too static and invariant a notion to do justice to the full complexity of the picture – not least since Sherman *et al.* do not refer to possible differences in police calls by *day* and *time* within their hot spot data?

A way to begin to answer these questions would seem to be to carry out observations of social activity in specific 'hot spots', observing both person–person interactions and the response of social actors to the physical aspects of location. Additionally, one could interview people who used these locations. Such proposed procedures, however, immediately raise questions about the way in which, as social scientists, we can make adequate sense of the relationship between statistical distributions (such as the Minneapolis 'hot spots' analysis) and ethnographic and interview studies of some of the same social phenomena. This is a problem which has bedevilled environmental criminology for a long time, and certainly since the days of the Chicago school. Unless and until it can be overcome, it seems likely that many aspects of place will go on being understated by general criminological writers, as they were by Braithwaite and by Gottfredson and Hirschi.

STATISTICAL DISTRIBUTIONS AND ETHNOGRAPHY: HISTORY AND PROBLEMATIC

The work of the pre-war Chicago researchers has justly remained important because they employed a wide variety of research methods, examining *inter alia* both the statistical data on offender distributions in the city (Shaw and McKay 1942) and aspects of the ethnography of street life and crime (e.g. Shaw 1930; Cressey 1932). They established that offender residence in Chicago was not randomly distributed across the city but was quite clearly patterned, with the highest offender rate areas located in an inner city zone close to the central business district, and then a diminution of the offender rate as one moved outwards towards the periphery of the city. In order to explain this distribution they utilized a theory of the growth of the city in terms of an historical process of urban development radiating outwards from the city's core. This theory seemed adequately to explain the distribution of land use they had found in Chicago, although it continued to be a matter of some debate as to how far the theory fitted other cities, and therefore whether it was a general theory. Their theory of urban development did not itself explain why offenders lived in some areas rather than others; however, from their ethnographic work the Chicagoans did develop an explanation of why and how offending occurred, based on the key concept of 'social disorganization'. Essentially their argument was that offending manifested itself in a lack of structurally located social bonds which encouraged legitimate and discouraged deviant behaviour. Such social disorganization was the result of new immigrant populations coming together and not having had the opportunity to develop a stable social structure with clear norms. Such populations were to be found in those areas of the city, immediately surrounding the inner core, which had been abandoned by more established groups and so offered the cheapest available housing for the new immigrants – the well-known 'interstitial areas' of the Chicago theory. The continuing process of immigration into Chicago meant that as immigrant groups developed more stable normative structures they moved out of the interstitial areas to be replaced, in their turn, by new immigrants. So the cycle was repeated, with new groups gradually developing from disorganization to more stable normative structures and at the same time moving their location gradually outward from the city's centre. In this way areas of the city continued to have patterned offender rates over time.

The Chicago theory of social disorganization has been very influential in the history of criminology. It appears to offer an answer to the problem of the relationship between studies of the areal statistical distribution of crime and offending, and studies of the ethnography of criminal behaviour. As a result much subsequent criminological research used the idea of social disorganization as a central concept. However, there are very real problems with the concept, and beginning with Whyte's (1943) classic *Street Corner Society*, it was subjected to a series of critiques. A number of writers pointed out that empirical studies of interstitial areas and/or deviant behaviour did not support the idea that illegal behaviour was always the result of 'disorganization' – rather, it might instead be the result of highly organized, but alternative sets of normative values. The fact that action is morally disapproved of does not mean that it is necessarily any less related to social organization (see e.g. Becker 1963). The result, it was argued, was that the theory, like a number of other social theories of crime, was overdeterministic and therefore over-predictive of crime (Matza, 1964). Basically the concept of social disorganization was attacked as being at best a value judgment, and at worst empirically false.

Although the concept of social disorganization has been subject to so much criticism it has nevertheless lived on. For example, recent discussion in the United States has used the notion of an 'underclass', whose lack of a normative order is said to be demonstrated by the collapse of the (black) family, to explain the high crime rates of their cities' ghettos (for a discussion of how these ideas have been used in popular debate see e.g. *Chicago Tribune* 1986). This renaissance of social disorganization is not entirely surprising since criminology has failed to develop any very satisfactory alternative concept to bridge the two levels of analysis. The alternative has all too often been simply to operate at just one level of analysis. Recent research in Britain has sometimes exemplified this approach. Janet Foster, in a study of crime on housing estates in south London, is most illuminating about the ethnography of crime but says little about the distribution of crime between or within estates (Foster 1990). On the other hand, the analysis of the results of the British Crime Survey examined the distribution of crime across socially different types of areas (using the ACORN classifications) but said little about why these areas have such different crime rates or indeed whether the classification which was used captured socially similar areas within its categories (Mayhew *et al.* 1989).[7] Of course, many problems of this kind may simply be due to the limitations of the particular research methods being used within a particular project, and in the end to the lack of the resources available to employ additional or alternative methods. However, the gap remains, and an adequate environmental criminology clearly needs a model of explanation which can link statistical analyses of the distribution of crime with ethnographic studies of criminal and social action.

STRUCTURATION THEORY

In order to explore what might be an adequate explanatory model for environmental criminology, we need first to explore the more general question of what an adequate explanatory model in social science might look like.

Social science has always had a problem with what form explanation ought to take, given that it is concerned with the activity of human beings. The twin dangers are that explanations either operate with models of human action which are so deterministic that they deny any role for human agency, or they are so voluntaristic and particularistic that they deny any real possibility of social science explanations at all. The history of social science could be written in terms of the various attempts to overcome this problem. Social science, like other human activities, has its fashions and at different times fashion has pushed researchers towards one or other of these extremes. The result has been that at different times explanations have been dominated by structural accounts, which have stressed the extent to which human behaviour is a product of the constraints imposed by social structures which are external to the individual (such as the economy), or alternatively by accounts of action, which have emphasized the extent to which human action is a consequence of the creative understanding of particular individuals, and their interaction with other actors. Both approaches have had the advantage of highlighting, often with great clarity, certain aspects of the human condition, but the disadvantage is that they remain partial.

Research in environmental criminology has been prone to just these difficulties. Explanations of where offences occur, or where offenders live, can all too easily assume that place or design acts as a deterministic and monocausal variable. Alternatively, they may assume that place can stand as an operational construct for other aspects of social structure, such as class, or employment status, or family structure; or that it is simply a sorting mechanism which brings together in one place those individuals who possess criminogenic attributes (generally of a genetic or psychological kind). These latter formulations use place as a second order explanation, parasitic on separate explanations of criminal behaviour, which simply accounts for the distribution of crime in geographical space. Alternatively again, structural explanations of this kind may combine the influence of place and the influence of class/employment/ family structure, yet remain straightforwardly deterministic. All of these approaches can be criticized as giving insufficient weight to human agency (see for example the work of Sally Merry (1981) on the limitations of a purely design-orientated approach to crime).

A very different approach has been the 'appreciative' one, which has produced a rich harvest of qualitative studies giving a vivid picture of life in a particular area, or the life history of individuals. The difficulty with this approach is that it can ignore the fact that there is a spatial patterning of crime which is in need of explanation, and/or that there are structural aspects in the wider society which powerfully shape the day-to-day lives of individual actors. One researcher who recognized just these difficulties was Owen Gill (1977), who set out to write an appreciative ethnography of the lives of a group of boys from a 'problem council estate' in Merseyside, but found himself successively drawn into social structural issues (and in particular the local housing market) in order to make adequate sense of his ethnographic data.

Writing on social theory and the methodology of the social sciences is replete with warnings against the partiality of 'structural' and 'action-based' approaches, and numerous attempts have been made, with varying degrees of success, to provide a framework for explanation which overcomes the problem. A particularly interesting recent approach to the issue has been made by Anthony Giddens in his 'theory of structuration', the very term combining the connotations of structure and action within a single theory (Giddens 1984). Giddens has argued not simply that explanations ought to be adequate at both these levels (as for example Max Weber did) but rather that it is a fundamental mistake to conceive of them as separate levels at all. Instead Giddens proposes that:

> The basic domain of study of the social sciences, according to the theory of structuration, is neither the experience of the individual actor, nor the existence of any form of societal totality, but social practices ordered across space and time.
>
> (Giddens 1984: 2)

Space and time are central to Giddens's model of explanation, and so it is especially appropriate to consider this approach in developing theories of environmental criminology. As Giddens puts the matter:

> [Most] social scientists have failed to construct their thinking around the modes in which social systems are constituted across time-space. ... investigation of this issue is one of the main tasks imposed by the 'problem of order' as conceptualized in the theory of structuration. It is not a particular type of 'area' of social science which can be pursued or discarded at will. *It is at the very heart of social theory, as*

interpreted through the notion of structuration, and should hence also be regarded as of very considerable importance for the conduct of empirical research in the social sciences.

(Giddens 1984: 110, italics added)

We shall not attempt to summarize all Giddens's arguments but merely to indicate the most important elements for our purposes. These include the following:

1 Human subjects are knowledgeable agents, though this know-ledgeability is bounded on the one hand by the unconscious, and on the other hand by unacknowledged conditions and/or unintended consequences of action.

2 Human subjects largely act within a domain of 'practical consciousness' which often cannot be expressed in terms such as 'motives' or 'reasons' but which 'consist of all the things which actors know tacitly about how to "go on" in the context of social life without being able to give them direct discursive expression' (Giddens 1984: xxiii). This 'practical consciousness' must, however, be understood and made plain by the researcher in explanation.

3 Structuration theory seeks to escape from the traditional dualism in social theory between 'objectivism' and 'subjectivism'. Thus the theory accepts concepts of 'structure' and 'constraint', normally associated with 'objective' social science, but insists that they be understood only through the actions of knowledgeable agents; on the other hand it believes that 'subjectivist' social science has overemphasized the degree to which everyday action is directly motivated.

4 Structures may act as constraints on individual action but they are also, and at the same time, the medium and outcome of the conduct they recursively organize – what Giddens refers to as the 'duality of structure'. Structures, therefore, do not exist outside of action, and they do not only constrain, but also enable social action.

5 'Routine' is a predominant form of agents' day-to-day activity: most daily practices are not directly motivated, and routinized practices are a prime expression of the 'duality of structure' in respect of the continuity of social life.

6 Structuration theory accepts and tries to elaborate Marx's famous dictum that human beings 'make history, but not in circumstances of their own choosing'. This is part of the duality of structure,

and emphasizes that social change and social process, linked to the reflexivity of human action, is an intrinsic part of human social life, even though that social life also has considerable continuities.

Giddens insists that both action and structure exist only within the ongoing process of human existence, which is largely constituted in practical consciousness. Structures, for Giddens, are properties which both allow and result in a practical consciousness which is able to follow regular patterns over time/space. The same practical rules which guide the social action of individuals are at the same time the basis for the reproduction of social systems. Looked at in this way the 'structure' of place is not simply a constraint on action but instead is one part of the social system which informs the practical (and sometimes discursive) consciousness of social actors. Put more simply, if we want to understand the geography of crime we have to understand how place, over time, is part of the practical consciousness of social actors who engage in behaviour, including actions we define as criminal. The structure of place is central, but it is not external to human agency and must be understood as part of an historical process.

Giddens's theory, then, gives us a model of explanation which we can use to examine critically some recent environmental criminology. Place cannot be made epiphenomenal to the explanation of human activity (as some human geographers once, suicidally, seemed to want to suggest) because place, together with time, are intrinsic dimensions of human existence. In acting, agents have to come to terms with the intrinsicality of space/time – which (as we have seen) they frequently do through routines. How they do so, whether they do so in different ways, and how modernity has extended the possible ways in which different actors operate in space/time are all interesting and empirical questions. All of us use our sense of 'locale' (for definition see note 6) to guide our everyday actions, and this is no less true in relation to crime. As Reiss put it, in mercifully straightforward terms,

> our sense of personal safety and potential victimisation by crime is shaped less by knowledge of specific criminals than it is by knowledge of dangerous and safe places and communities.
>
> (Reiss 1986: 1)

To this one might add, first, that the general public's sense of safety relates not only to place but also to different times of day in place, and second, that the everyday life of offenders, as well as of victims and potential victims, is shaped in part by understandings of the nature of particular areas and, within them, of specific locations – and those understandings are undoubtedly important in shaping the geographical distribution of offending behaviour.

STRUCTURATION AND ENVIRONMENTAL CRIMINOLOGY

Let us now consider how structuration theory can help take forward the study of environmental criminology.

To assess this issue we shall examine a recent essay by Per-Olof Wikström (1990), written in an attempt to summarize the literature on crime, criminality and the urban structure as a background paper for a new and major empirical research project in Stockholm. Wikström's paper is both up-to-date and of high quality; it provides, therefore, a useful exemplar of the 'state of the art' in environmental criminology, and a way of testing whether the application of a structuration approach (which Wikström does not consider) might have something to offer to this field of study.

Like most environmental criminologists, Wikström draws a clear distinction between area offender rates and area offence rates (see note 3). In summarizing the relationship between urban structure (especially housing) and area offender rates, Wikström postulates two main effects:

1 Housing and [offender-rate based] criminality are related because social groups with a greater propensity to crime are concentrated in certain types of housing. ...
2 Housing can itself affect the resident's propensity to crime in that the local housing conditions are of importance both to the social life and the social control of the neighbourhood (the 'contextual' effect). This effect may be subdivided into
  (a) situational influence on propensity to offend; and
  (b) long-term influence on the development of the individual resident's personality and life-style, tending to reinforce a propensity to crime ... (primarily applies to neighbourhood influences on children and young people).

(Wikström 1990: 17)

This summary is in many ways extremely closely related to the approach we have ourselves developed from empirical work in Sheffield, which is discussed in Chapter 7 in this volume.

Turning to the relationship between urban structures and area offence rates, Wikström adopts an approach arising out of routine activities theory, and opportunity theory:

> Inter-district variations in the use of urban land generate different activities more or less frequently *at different times of the week and day in different parts of the city*. Segregation and the spatial variation in the pursuit of various activities, each of which will be perceived as more or less attractive by different social groups, ensure that *the social make-up of residents and visitors at different times of day will show distinct inter-district variations*.
>
> The type of activities being pursued and the social com-position of the people in the district at any one time can be assumed to be related to
>
> 1 the availability of suitable criminal targets, the presence of motivated offenders and the presence of direct social control (capable guardians) [explanation of offence rates for instrumental crime]
> 2 the occurrence of encounters (environments) liable to provoke friction in the parochial and public orders [explanation of some expressive crime].
>
> (Wikström 1990: 23, italics added)

Wikström then offers a diagram (reproduced here as Figure 1.1) summarizing his approach to the explanation of offence rates.

These summary statements undoubtedly capture much of our present knowledge about the reasons for inter-area variation in offender rates and offence rates within cities.[8] They are a bold and interesting attempt at synthesis, though – as Wikström would, we think, be the first to agree – they incorporate within them both some points with solid support in empirical research, and others which in the present state of knowledge seem reasonable or even probable, but where the empirical support is much more slender.

From the perspective of structuration theory, two points stand out as interesting in Wikström's summaries. The first concerns the marked emphasis, in the offence rate summary, on the differential paths taken by different actors in space and time, linking with

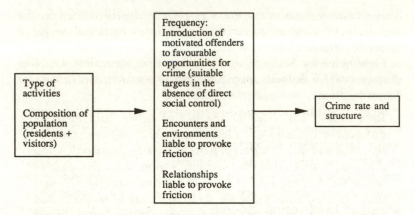

Figure 1.1 Variation in crime (offence rate) and structure in the urban environment (after Wikström 1990)

structuration theory's emphasis on space/time issues. The second and closely related point concerns the use of Cohen and Felson's routine activities theory, emphasizing the extent to which offences either arise directly out of the routine (legal) activities of social actors, or how even a deliberately and consciously illegal activity (e.g. a planned burglary trip) may in practice be confined to areas or trunk roads already known to the offender(s) through their everyday lives (see Brantingham and Brantingham 1981). Clearly, there is an intriguing link to be forged here between the centrality of *routines* in Giddens's structuration theory (itself an interesting innovation in general sociological theory) and the emerging importance of routine activities theory in environmental criminology.

But while one can see important points of contact between structuration theory and Wikström's approach, a structuration perspective also suggests some weaknesses in his otherwise excellent paper. Two such weaknesses seem particularly apparent, and it is worth elaborating these by way of constructive criticism.

First, Wikström's approach markedly understates the importance of social process. In his offender-rate based analysis, for example, it can be argued that insufficient attention is paid to the constantly changing nature of the local housing market, in particular districts (often linked to more macro-level economic changes, or alterations in government housing policies). In structuration theory, structures are always simultaneously both enabling and constraining, and therefore never static. Similarly, Wikström's diagram for considering

22

offence rate variations (Figure 1.1) gives inadequate attention to the changes (as opposed to the continuities) in the use by social actors of different districts within the city, or (at a more micro-level as highlighted by Sherman *et al.* 1989) to the constantly evolving character of, for example, particular streets or particular bars in city centre locations.

Second, Wikström's approach arguably pays too little attention to the perceptions, routine activities, and decisions of individual actors (as opposed to aggregate patterns of social activity). In so doing, he risks missing the important distinction, central to structuration theory, between the intended and unintended consequences of individual action. Yet unintended consequences of action can be of central importance in, for example, the operation of housing markets (itself central to Wikström's framework for understanding offender rate variations); or the evolution of particular micro-level locations towards or away from being 'hot spots' of crime.

A more detailed consideration of the issues of process and unintended consequences may therefore help in the development of a more adequate approach to environmental criminology.

UNINTENDED CONSEQUENCES OF ACTION AND PROCESSES OF CHANGE

An important aspect of Giddens's structuration theory is that the structures which result from human action are not just a result of the intended consequences of such actions. Indeed, Giddens's notion that human action largely follows 'routines' precisely emphasizes the inadequacy of a fully intentional model of human action. The result is that not only may the consequences of rationally calculated action be unforeseen by actors (or foreseen but unintended), but also human action may frequently not be guided by conscious intention at all. In sum, the structural results of a series of human actions may be quite different from what the actors may have foreseen or intended.

Giddens gives as an example a model of how racial segregation in a city might occur:

A pattern of ethnic segregation might develop, without any of those involved intending this to happen, in the following way, which can be illustrated by analogy. Imagine a chessboard which has a set of 5-pence pieces and a set of 10-pence pieces. These are distributed randomly on the board, as individuals might be in an urban area. It is presumed that, while they feel no hostility towards the other group, the members of each group do not want to live in a neighbourhood where they are ethnically in a minority. On the chessboard each piece is moved around until it is in such a position that at least 50 per cent of the adjoining pieces are of the same type. The result is a pattern of extreme segregation. The 10-pence pieces end up as a sort of ghetto in the midst of the 5-pence pieces. The 'composition effect' is an outcome of an aggregate of acts – whether those of moving pieces on the board or those of agents in a housing market – each of which is intentionally carried out. But the eventual outcome is neither intended nor desired by anyone. It is, as it were, everyone's doing and no one's.

(Giddens 1984: 10)

It is interesting that Giddens introduces the concept of a 'market' into this example, since this concept is used by economists to signify the summation of the consequences of individual economic decisions, regardless of whether those consequences were intended or foreseen by the actors. Taub *et al.* (1984), in their study of the decline of neighbourhoods in Chicago in both racial and crime terms, use a similar market-based model to explain the actions of individual householders (as opposed to corporations, etc). They argue that such individual residents, when faced with signs of neighbourhood decline, can take decisions only in terms of their own purposes and in the context of their (limited) understanding of what other residents will do. The result can be that, in a similar way to Giddens's example, while none of the residents have an interest in the neighbourhood declining, the unintended consequence of their individual decisions can be precisely that.

The idea of a 'market', of course, is a model developed to help understand the aggregate results of action from an economic point of view and although, as Taub *et al.* have shown, it can be usefully employed to help explain the processes by which neighbourhoods decline it is not in itself wholly adequate to explain why a neighbourhood's crime or offender-rate pattern changes. What seems to be needed, then, is a model of neighbourhood activity which helps us to understand how changing crime or offending behaviour can be the result of a summation of individual actions, and their intended, unintended and unforeseen consequences: in other words, we need a construct which will fulfil for criminology some of the functions which the market fulfils for economists.

Some time ago Albert Reiss suggested that changes in neighbourhood crime patterns could be thought of as analogous to communities having crime careers (Reiss 1986). Reiss did not develop this idea of 'community crime careers' much further, but in our view it is extremely suggestive. The term 'community crime career' in effect encompasses the notion that a neighbourhood's crime pattern is the summation of the consequences, whether intended or not, of the way a multitude of actors interact (which itself is linked to their practical consciousness of locale) in an historical process. As such it can equally be applied to offender changes in or offence-rate crime patterns, or to the relationship between the two. We found Reiss's idea so useful that we have employed it in some of our recent analyses of the Sheffield research (Bottoms and Wiles 1986; see also Chapter 7 in this volume) but we are aware that the concept needs further development if its full potential is to be realized.

Some help in developing this concept may perhaps be obtained by considering again the work of Taub *et al.* (1984), although the writing of these authors predates that of Reiss. When they constructed a general theory of neighbourhood change out of their research, Taub *et al.* argued that

> There are three types of social and ecological pressures that interactively determine the pattern of change in urban neighborhoods: (1) ecological facts; (2) corporate and institutional decisions; and (3) decisions of individual neighborhood residents.
>
> (Taub *et al.* 1984: 182)

They pointed out that, traditionally, most urban theorists have concentrated on the 'ecological facts'[9] as the main explanatory variable as regards general social change in neighbourhoods, giving a strongly structural quality to such explanations. Such an emphasis, Taub and his colleagues believed,

> gives the wrong impression about the dynamics of neighborhood change. Individual residents and local corporate actors are, after all, the ones whose day-to-day decisions define the texture and quality of urban life. If ecological facts are

overwhelming, it is because of the effect of these facts on the perceptions and actions of individual and corporate actors. In a neighborhood that goes up or down, it is ultimately the actions of these residents that make the outcomes real.

(Taub et al. 1984: 186)

Although their language is different these authors are essentially following the model of explanation proposed by Giddens in insisting that 'place' has to be considered always as it is constituted through human action. Their threefold interactive model of ecological facts/individual decisions/corporate decisions also offers a valuable framework for analysis of area change,[10] even though the concept of 'ecological facts' needs some reinterpretation from the standpoint of structuration theory.[11]

Despite its merits, however, Taub et al.'s model is of limited value for present purposes because the analysis of individual and corporate actors' perceptions and decisions is applied only to the operations of the property market and its consequences. If we are to develop the notion of 'community crime career' more generally, then we need to extend this type of analysis to encompass all the structures which are relevant to the processes of change in offender and offence rates, as illustrated, for example, in Wikström's summaries. We return to this point, with examples, in the following section.

One other matter, of some importance for what we would regard as an adequate development of Taub et al.'s analysis, must be raised here. Even where actors appear to be operating in a market type situation, a model which focuses on their actions as motivated solely towards that market will have serious inadequacies. One only has to reflect momentarily on the reality of actors' behaviour in parts of the British housing market to see why this is so. Even after the changes of the 1980s a significant proportion of the British housing stock exists within a market of bureaucratic allocation, whose rules are very different from those of a price market. (For details of how this operates in one city, see Chapter 7 in this volume.) In this sector actors need a very sophisticated understanding of the rules of the market in order adequately to foresee the consequences of their choices; yet there is some evidence to suggest that in some areas the allocation process is operated by the local authority in such a paternalistic way that actors do not even perceive that they have a choice. Even where actors do understand that they have a choice, and understand the rules of allocation, it does not necessarily follow that they will maximize the benefits available to them as the model of a rationally calculating market actor would suggest. In the 1980s replication stage of the Sheffield research tenants in a notorious high-rise block of flats had to be re-housed due to its impending demolition. Because of the rules of allocation in Sheffield, these 'clearance' tenants had priority in the allocation to vacant housing units in the local authority's stock, and they could therefore have secured transfers on to some of the most select council estates in the city. In fact few of them chose to do this even though advisers, ranging from community workers to police community constables, explained to them how to do so. Most of them instead chose to move to nearby (and by no means select) estates. The reason was because their bounded sense of location meant that they regarded the alternative select estates as inappropriate for them either in geographical terms (they did not 'belong' in a different sector of the city) or in class terms (the select estates were for the 'respectable'). Such a sense of locale is particularly powerful in Sheffield: one community constable, who was bemoaning the tenants' refusal to maximize the advantage available to them, was reminded by his colleague that he had himself declined a transfer to a different police division because he didn't feel at home in the area! However, a sense of location is merely one of the other aspects of structure which needs to be built in to the full development of a concept of community crime career, as should become clear from the final section of this chapter.

DEVELOPING THE COMMUNITY CRIME CAREER CONCEPT: OFFENDER AND OFFENCE RATE VARIATIONS RECONSIDERED

The concept of a community crime career is, of course, a concept embodying the idea of social change at a meso-level. Structuration theory, as we have seen, places considerable emphasis on social process and social change, holding indeed that such matters are intrinsic to social life as lived out by human agents acting reflexively. But Giddens (1984: ch. 5) also argues, correctly in our view, that given the premises of structuration theory no general theory of social change is possible:

> The reflexive nature of human social life subverts the explica-
> tion of social change in terms of any simple and sovereign set

of causal mechanisms. ... To insist that social change be studied in 'world time' [i.e. examination of all social conjunctures in the light of reflexively monitored 'history'] is to emphasise the influence of varying forms of intersocietal system upon episodic transactions. If all social life is contingent, all social change is conjunctural. That is to say, it depends upon conjunctions of circumstances and events that may differ in nature according to variations of context, where context (as always) involves the reflexive monitoring by the agents involved of the conditions in which they 'make history'.

(Giddens 1984: 237, 245)

This does not mean, as Giddens goes on to emphasize (1984: 244f) that we cannot generalize at all about social change; it does mean, however, that there 'are no universal laws in the social sciences, and there will not be any' (1984: xxxii).

The implications of this for the concept of community crime careers are, first, that exact prediction of the precise course of such careers will be impossible, but second, that we should be able to analyse with some precision the general factors that may influence the development of such careers, in a probabilistic manner. At the present time our ability to specify these factors is rather rudimentary, but it should improve if more scholars develop empirical research specifically based upon the concept of community crime careers.

Let us take two particular examples, one focused upon an *offender-rate based* community crime career, and the other on an *offence-rate based* career. As to the former, Wikström's formulation (discussed earlier) carries considerable plausibility, provided that it is supplemented by an understanding of processes of social change and of the unintended consequences of choices made by individuals and corporate actors within the housing market, as previously discussed (see also Chapter 7 in this volume). Wikström highlights the allocative functions of the housing market, and the contextual effect of housing allocations (itself divided into 'situational' and longer-term effects) as the explanation of changing offender rates. A way of elaborating the issues relevant to this approach, modifying an idea originally suggested by Wikström's colleague Peter Martens (1990: 66), is shown in Figure 1.2. In considering this diagram, it should be noted that in our view the housing market context (on the left of the diagram) has to be especially prioritised in the explanation, for the reason that it is the housing market which is

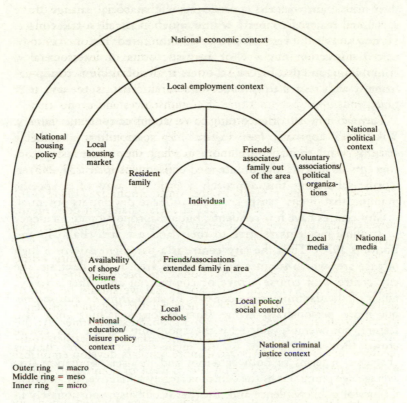

National economic context

Local employment context

National housing policy

Local housing market

Resident family

Friends/ associates/ family out of the area

Voluntary associations/ political organiza- tions

National political context

Individual

Local media

National media

Availability of shops/ leisure outlets

Friends/associations extended family in area

Local schools

Local police/ social control

National education/ leisure policy context

National criminal justice context

Outer ring = macro
Middle ring = meso
Inner ring = micro

Figure 1.2   Heuristic model of the context of offending

responsible for the allocation of families and individuals to particular areas; but all the other matters shown in the diagram also come into play by way of contextual effects (or, as we put it in Chapter 7, indirect effects).

Figure 1.2 explicitly employs the terminology of 'macro', 'meso' and 'micro' processes. We use these terms for heuristic purposes only. We are aware that Giddens (1984: 139–44) eschews them and insists that such distinctions are merely interconnected aspects of how the social phenomenon being examined is located in space/ time. Certainly Giddens is right to make this point, and we agree with him that structure is as relevant to micro-sociology as to macro-sociological issues, while no macro-sociological structure can be adequately understood aside from an understanding of the purposive decisions and routine activities of human agents, and their interaction with the constraining and enabling features of the social and material contexts within which that action takes place. Having said that, it remains in our view the case that a micro/meso/ macro distinction has a clear heuristic value in environmental criminology, and that Figure 1.2 offers a possible way of conceptualizing it with regard to offender-rate variations in space and over time.

Turning now to offence-rate based community crime careers, Wikström's approach (see Figure 1.1) again offers a valuable starting-point, though once more needing supplementation from concepts of process and unintended consequences. It is, however, worth elaborating this approach a little by way of a specific example, that of city centre crime, in order to develop its potential.

City centres have few residents, but a disproportionate incidence of criminal incidents relative to their land use area (Baldwin and Bottoms 1976). Thus the city centre is a paradigm case of a high offence area which is not a high offender area. In the daytime, the city centre is of course a hive of commercial and other activity; much of the daytime crime consists of shoplifting or auto-crime, but more personalized crime such as bag-snatching may also feature, considerably aided by the anonymity of the city centre crowd (Poyner 1983: ch. 6). At night the city centre changes character in terms of both activities and its user population (the average age plummets as the centre is largely taken over by youths). Crimes of public violence and disorder occur disproportionately in city centres at night, often in or close to pubs, clubs or other places of entertainment: these crimes are highly focused upon Friday and Saturday evenings, with a special time focus on pub and club closing times (Hope 1985; Ramsay 1982; Wikström 1985). Such incidents are also highly localized, with a few locations providing a disproportionate share of the crimes (Hope 1985; Sherman et al. 1989).

We immediately see here clear evidence of differential social activity in space/time by different groups, even in an area which is open to all. Further detailed research would probably show age, class and/or sex segregation in specific locations within the city centre both by day and by night – and would almost certainly show some city centre users to be anxious about groups of youths hanging about (Phillips and Cochrane 1988) or vagrant alcoholics (Ramsay 1989). Owners of specific premises may seek to achieve a degree of social segregation by manipulating the sense of location (elaborate entrance portals to an exclusive hotel; interior design deliberately calculated to make a particular age group, class or sex feel at home); others may seek to boost a sense of safety through the employment of private security companies (as, increasingly, in shopping malls with multiple retail outlets) or physical security devices (use of CCTV, and so on). There is clearly here a rich field of exploration in the patterning of use of the city centre, and, by those who do use it, in the patterning of the use of specific sites. Nor is any of this static. Perceptions of the desirability of a particular shop/café/bar can easily change over time, and individual decisions about custom can cumulatively have important long-term consequences. Add to this the fact that city centres themselves change over time, both in their land use and design (increasing use of pedestrianized streets, increasing development of multiple stores rather than small shops, and so on) and in their social use (the city centre in the evening in the early 1960s was much less predominantly a young person's domain), and we begin to see the complexity of the whole picture within which city centre crime must be understood. Wikström's model (Figure 1.1) is correct in so far as it goes, but needs a much greater understanding of the fluidity of social routines and practices before it can be fully adequate. That is true also of Sherman *et al.*'s 'hotspots', where specific observations and interviews need to be developed, enabling the researcher to understand why this particular location is used in this particular way, how it attracts or enables (particular kinds of) crime to take place, and how it changes over time.

These examples are no more than suggestive, and the concept of a community crime career clearly requires further elaboration. We hope, however, that the examples may help to bring alive some of the theoretical issues discussed in more abstract terms earlier in this chapter.

CONCLUSION

What we have tried to argue in this chapter is that a proper understanding of the spatial aspects of offences and offending is possible only if a model is employed which is capable of including the natural and built environment, the political, economic, social and cultural contexts and structures of areas and the actions of individuals and corporate bodies within areas, within a theory which accounts for the ongoing processes of interaction between them. We have used Giddens's structuration theory because we believe that it offers such a framework, and we have tried to show how the elements of this approach can illuminate current work within environmental criminology.

The task of developing a more adequate environmental criminology seems to us rather urgent at the present time, since for a variety of reasons the structures and understandings which underpin much existing scholarship are undergoing some rapid changes. For example, the British housing market context (which we discuss in detail in Chapter 7) has in fact already changed since we completed our latest empirical research. The tenure map of Britain is being redrawn, and this may well have significant consequences for the geography of crime. Perhaps more fundamental is the movement of the urban middle class into rural communities, and the consequential push of the less affluent rural young into the towns in search of affordable housing: the old rural/urban crime patterns will amost certainly be affected by this process. In the towns themselves, city centres are increasingly facing competition from large out-of-town shopping areas on the North American model, where rather different strategies of social control and segregation are being deployed. Industry and commerce, and even retailing, are also increasingly being segregated from residential areas and placed on industrial estates, technology parks, and so on. At the same time as the geography of Britain is being altered in these ways, so also technological innovations are constantly affecting the ability of social actors (both individuals and corporate bodies) to manipulate the constraints of both time and distance. If, as Giddens suggests, our sense of location is a key aspect of our social existence and is a product of our experience and interaction within space/time then the sense of location of many modern Britons is likely to be significantly changed. If it is, then the appropriateness of place to life-style, including the criminal, and to life experiences, including victimization, will also change. It is therefore vital that we should possess an adequate model for understanding these processes and their criminological consequences. We can at any rate confidently predict that there will be no shortage of interesting research topics for future environmental criminologists.

NOTES

1 Gottfredson and Hirschi are the more ambitious: they claim 'to explain all crime, at all times, and, for that matter, many forms of behaviour that are not sanctioned by the state' (1990: 117). Braithwaite describes his theory as 'a general theory of crime' (1989: vii), but restricts his attention to economically advanced societies and specifically excludes from the theory 'the small minority of criminal laws that are not consensually regarded as justified' (1989: 3).

2 Though it is not without its problems, especially in an era of 'green' issues, we use this concept throughout this chapter as the most convenient generic description of analyses of the relationship between place, crime and offending. Other such descriptions include 'the geography of crime', 'the ecology of crime', etc.

3 For the offence/offender distinction, see the discussion of Wikström's work on pp. 20–3; for a full consideration of the matter, see Bottoms and Wiles (1986: 158–60).

4 A very relevant issue in elaborating these themes would be the distinction drawn by Rhodes and Conly (1981) between 'target attractiveness' and 'spatial attractiveness'.

5 That is, all telephone calls for police assistance excluding non-crime calls such as calls for ambulances, fire service, etc.

6 Sherman *et al.* define 'place' as 'a fixed physical environment that can be seen completely and simultaneously, at least on its surface, by one's naked eyes' (1989: 31). Our own usage in this chapter, not least in the title, is of course considerably broader than this. It should be added, however, that Sherman *et al.* are not blind to what they call the 'sociological concept of place', defined as 'the social organization of behaviour at a geographic place'. This point is developed and strengthened by Giddens (1984) in his concept of 'locales', which, he insists, 'are not just places but *settings* of interaction, the settings of interaction in turn being essential to specifying its *contextuality*' (pp. xxv, 118). Giddens's formal definition of a locale is as follows: 'a physical region involved as part of the setting of interaction, having definite boundaries which help concentrate interaction in one way or another' (1984: 375).

7 For example, as far as we can tell, both of the areas discussed in Chapter 7 in this volume (Stonewall and Gardenia) would, despite their very different cultural histories, fall into ACORN Group F (Mayhew *et al.* 1989: 108).

8 Even within its own terms, however, it can be argued that Wikström's formulation pays too little attention to the implications of spatial form and design for offence rate distribution (see Newman 1972). Although we are stressing (following Giddens) that it is actors' conceptions of 'locale' which are critical (either in direct awareness or more usually in practical consciousness), nevertheless the physical nature of the environment must place some limits on what these conceptions can consist of.

9 'Ecological facts', according to Taub *et al.*, 'define the social and economic context for a neighborhood' (1984: 182). The ecological

facts of particular importance for neighbourhood decline are said to be:

1 the potential employment base for neighbourhood residents
2 demographic pressures on the neighbourhood housing market
3 the age and original quality of the housing stock
4 external amenities such as attractive physical locations (hills, views, and so on).

10 Taub et al.'s distinction between corporate and individual actors is important because it recognizes that in a market situation corporate actors have greater power to influence outcomes, whether such corporate actors are commercial corporations, non-commercial organizations such as universities (the University of Chicago features strongly in one of Taub et al.'s area case studies) or individual actors who band together in a political or community organization in order to derive the benefits of corporate power. Of course, in markets the ultimate corporate actors are either those who hold a monopoly, or those who can use the legislative and administrative power of the state to redefine the nature of the market.

11 Some of the 'ecological facts' as defined by Taub et al. (e.g. the local employment base) are of course themselves part of the social structures which are constituted and reproduced in human action.

REFERENCES

Baldwin, J. and Bottoms, A.E. (1976) *The Urban Criminal*, London: Tavistock.

Becker, H.S. (1963) *Outsiders*, New York: Free Press.

Bottoms, A.E. and Wiles, P. (1986) 'Housing tenure and residential community crime careers in Britain', in A.J. Reiss and M. Tonry (eds) *Communities and Crime*, Chicago: University of Chicago Press.

Braithwaite, J. (1989) *Crime, Shame and Reintegration*, Cambridge: Cambridge University Press.

Brantingham, P.L. and Brantingham, P.J. (1981) 'Notes on the geometry of crime', in P.J. Brantingham and P.L. Brantingham (eds) *Environmental Criminology*, Beverly Hills, Calif.: Sage.

Chicago Tribune (eds) (1986) *The American Millstone*, Chicago: Contemporary Books.

Cohen, L.E. and Felson, M. (1979) 'Social change and crime rate trends: a routine activities approach', *American Sociological Review* 44: 588–608.

Cressey, P.G. (1932) *The Taxi-Dance Hall*, Chicago: University of Chicago Press.

Davidson, N. (1981) *Crime and Environment*, London: Croom Helm.

Foster, J. (1990) *Villains: Crime and Community in the Inner City*, London: Routledge.

Giddens, A. (1984) *The Constitution of Society*, Cambridge: Polity.

Gill, O. (1977) *Luke Street*, London: Macmillan.

Gottfredson, M.R. and Hirschi, T. (1990) *A General Theory of Crime*, Stanford: University Press.

Gregory, D. and Urry, J. (1985) *Social Relations and Spatial Structures*, London: Macmillan.

Herbert, D.T. (1982) *The Geography of Urban Crime*, London: Longman.

Hope, T. (1985) *Implementing Crime Prevention Measures*, Home Office Research Study 86, London: HMSO.

Martens, P.L. (1990) 'Family, neighbourhood and socialisation', in P.-O.H. Wikström (ed.) *Crime and Measures against Crime in the City*, Stockholm: National Council for Crime Prevention.

Matza, D. (1964) *Delinquency and Drift*, London: Wiley.

Mayhew, P., Elliott, D, and Dowds, L. (1989) *The 1988 British Crime Survey*, Home Office Research Study 111, London: HMSO.

Merry, S. (1981) 'Defensible space undefended: social factors in crime prevention through environmental design', *Urban Affairs Quarterly* 16: 397–422.

Newman, O. (1972) *Defensible Space*, New York: Macmillan.

Phillips, S. and Cochrane, R. (1988) *Crime and Nuisance in the Shopping Centre*, Crime Prevention Unit Paper 16, London: Home Office.

Poyner, B. (1983) *Design Against Crime: Beyond Defensible Space*, London: Butterworths.

Ramsay, M. (1982) *City Centre Crime*, Home Office Research and Planning Unit Paper 10, London: Home Office.

Ramsay, M. (1989) *Downtown Drinkers: The Perceptions and Fears of the Public in a City Centre*, Crime Prevention Unit Paper 19, London: Home Office.

Reiss, A.J. (1986) 'Why are communities important in understanding crime?' in A.J. Reiss and M. Tonry (eds) *Communities and Crime*, Chicago: University of Chicago Press.

Rhodes, W.M. and Conly, C. (1981) 'Crime and mobility: an empirical study', in P.J. Brantingham and P.L. Brantingham (eds) *Environmental Criminology*, Beverly Hills, Calif.: Sage.

Shaw, C.R. (1930) *The Jack Roller*, Chicago: University of Chicago Press.

Shaw, C.R. and McKay, H.D. (1942) *Juvenile Delinquency and Urban Areas*, Chicago: University of Chicago Press.

Sherman, L.W., Gartin, P.R. and Buerger, M.E. (1989) 'Hot spots of predatory crime: routine activities and the criminology of place', *Criminology* 27: 27–55.

Smith, S.J. (1986) *Crime, Space and Society*, Cambridge: Cambridge University Press.

Taub, R., Taylor, D.G. and Dunham, J.D. (1984) *Paths of Neighborhood Change*, Chicago: University of Chicago Press.

Whyte, W.H. (1943) *Street Corner Society*, Chicago: University of Chicago Press.

Wikström, P.-O.H. (1985) *Everyday Violence in Contemporary Sweden*, Stockholm: National Council for Crime Prevention.

Wikström, P.-O.H. (1990) 'Delinquency and urban structure', in P.-O.H. Wikström (ed.) *Crime and Measures against Crime in the City*, Stockholm: National Council for Crime Prevention.

2

LEFT REALISM AND THE SPATIAL STUDY OF CRIME

David J. Evans

This chapter focuses upon certain of the propositions outlined by Lea and Young in their left realist position in criminology. Left realism has a number of features which distinguish its position from that of the preceding criminological position, namely that of left idealism. These are its focus upon working-class crime or what may be called 'normal crime' in the sense that there is a readily identifiable victim and that it is these crimes which appear most frequently in police statistics and are of greatest concern to the public. The focus of left idealism was upon the crimes of the powerful, the effects of which may be more pervasive but which are nevertheless more invisible and less evident to the public. Such an emphasis by left idealists meant that left idealism was concerned with theory at the macro-scale, whereas left realism studies crime from both the macro and micro perspectives. This means that left realism is more comprehensive and that it has a greater empirical content (see Young 1987: 337). Left realism has also rediscovered the victims of crime and placed the emphasis upon the victims and their experiences and fears rather than placing the emphasis upon the offenders and their roles as 'victims'. A third characteristic of left realism is its greater development of empirical enquiry especially through the use of victimization surveys. Official crime statistics are seen as subjective and political (Lea and Young 1984: 16) and are not therefore accurate reflections of the incidence of crime.

The practices and assertions of left realism are examined in relation to available empirical information which is drawn in the main from studies conducted in the United Kingdom. The first issue addressed is the use of victimization surveys as an alternative to official crime statistics. Such surveys conducted usually at a local scale (in, for example, Islington, Merseyside, and Hammersmith and Fulham) have become an important source of information about crime in specific localities. They do, however, have limitations which should be examined so that their efficacy can be assessed. The second assertion which is examined is that 'the working class is a victim of crime from all directions; that one sort of crime tends to compound another' (Lea and Young 1984: 264). Lea and Young were thinking here that the working class were victims of crimes of the powerful and victims of crimes from other members of the working class. It does, however, also raise the issue of multiple victimization, a feature of crime neglected by social geographers but a characteristic which logically follows from concepts of geographical and social risk. It follows because vulnerable locations and people are more likely to be victimized not once but a number of times. A further issue to be examined in this context is the appropriate categories to use in victimization surveys. Left realists (Lea and Young 1984: 31) argue that the appropriate level of analysis focuses within social categories rather than using broad social categories. Young (1987: 348) argues that crime studies should use 'precise groupings of people according to the major social axes: gender, age, class and race, which form the basis of actual lived subcultures'. In this way greater precision in terms of victimization is obtained and some of the apparent 'irrationalities' in fear of crime studies can be explained. The third proposition to be analysed is that 'fear of street crime helps the disintegration of the working-class community' (Lea and Young 1984: 264). The fourth area of discussion is the left realist position with regard to the causes of crime. This position is both macro and micro in nature. Spatial studies can add micro detail and through the use of matched pairs show the different offender rates of areas with similar social characteristics. Finally the statements made by left realists that 'environmental and public precautions against crime ... do not relate to the heart of the matter' and that 'the organisation of communities in an attempt to pre-empt crime is one of the utmost importance' (Lea and Young 1984: 269) are also examined. How effective is the neighbourhood watch movement in preventing crime? Is it more effective than alternative methods?

PROPOSITION ONE: VICTIMIZATION SURVEYS SHOULD BE USED IN PREFERENCE TO OFFICIAL STATISTICS

Lea and Young (1984), with regard to official crime data, state that

we have pointed out the subjective and political nature of the 'official' crime statistics. This is not to suggest that the crime problem lacks reality; far from it. It is to say that we must handle the figures with caution, and most importantly that we must develop a sense of realism.

(Lea and Young 1984: 16)

Part of this realism is the use of victimization surveys as an indicator of 'true' crime levels. Before specifying the problems of victimization surveys, the reporting and recording levels of residential burglary and of violent crimes will be examined to illustrate the problems of officially derived data on crime.

Official crime statistics

Official crime statistics suffer from at least two problems, namely those of differential reporting and differential recording levels for different crimes. The respective reporting and recording levels of residential burglary and of violent crime illustrate the problems involved to different degrees. Table 2.1 shows the reporting rates for residential burglary and violent offences as revealed by a series of national and local victimization surveys. The average reporting rate for burglary is 68.1 per cent and 54.5 per cent, 42.3 per cent, 30.7 per cent and 20.2 per cent for robbery, wounding, assaults and sexual offences respectively.

As important as the rate of reporting is the influence of age, social class, sex, ethnicity and attitude to the police of the victim upon reporting behaviour. The second British Crime Survey (Hough and Mayhew 1985: 24, 68) suggests statistically significant variations in reporting behaviour of burglary cases by social class but not by age, sex, education level an attitudes to the police. The General Household Survey (OPCS 1982: 81) also shows differential reporting behaviour by social class. With regard to ethnicity, the studies by Smith (1982) and Tuck and Southgate (1981) show conflicting results. The former suggests that for all crimes Asian respondents report at higher levels than West Indian respondents; the respective

Table 2.1 Reporting rates for residential burglary and violent crime

Residential burglary	%
First British Crime Survey	66 (Hough and Mayhew 1983: 11)
Second British Crime Survey	68 (Hough and Mayhew 1985: 61)
Third British Crime Survey	63 (Mayhew *et al.* 1989: 16)
General Household Survey	78 (Burglary and theft from private households) (OPCS 1982: 81)
Sheffield survey	68 (Bottoms *et al.* 1987: 145)
London survey	51 (Burglary and theft in a dwelling) (Sparks *et al.* 1977: 157)
Islington Crime Survey	78 (MacLean *et al.* 1986: Table 3.4)
Merseyside Crime Survey	73 (Kinsey 1984: 11)

Violent Crime		%
First British Crime Survey	Wounding	39
	Robbery	47
	Sexual offences	28
(Hough and Mayhew 1983: 11)		
Second British Crime Survey	Wounding	60
	Robbery	57
	Common assult	31
	Sexual offences	10
(Hough and Mayhew 1985: 21)		
Third British Crime Survey	Wounding	43
	Robbery	44
	Common assault	33
	Sexual offences	21
(Mayhew *et al.* 1989: 16)		
Islington Crime Survey	All assaults	38.4
	All sexual assaults	21.9
	Robbery	52.6
(MacLean *et al.* 1986: Table 3.4)		
Sheffield survey	Assaults and threats	22
(Bottoms *et al.* 1987:145)		
Merseyside Crime Survey	Wounding	27
	Robbery	72
(Kinsey 1984: 11)		
London survey	Assault/theft from person	29
(Sparks *et al.* 1977: 157)		

percentages being 53.8 per cent, 35.3 per cent, 21.2 per cent of all crime incidents. The Tuck and Southgate study reverses these levels for white and West Indian respondents for break-ins, the respective reporting levels being 68 per cent and 80 per cent.

The influence of ethnic background upon the reporting of burglary cases is not clear at least from the results of these two studies. Social class appears to be the most consistent factor influencing reporting of burglary cases.

By contrast the second British Crime Survey showed that while serious crimes of violence have statistically significant variations in reporting by sex only, less serious crimes of violence have statistically significant variations in reporting levels by class, age and sex (Hough and Mayhew 1985: 24, 67).

Police recording levels for burglary and violent offences are shown in Table 2.2; the average recording level for burglary is 62.2 per cent and 46.3 per cent, 27.2 per cent and 84.3 per cent for robbery, assault and sexual offences. This suggests that the recording of violent offences is more problematical than the recording of burglary offences (see, for example, Bottomley and Coleman 1981: 61). The main reason for the 'no-criming' of crimes of violence in the Bottomley and Coleman study was the victim's wishes (Bottomley and Coleman 1981: 73).

Table 2.2 Recording levels for residential burglary and violent crime

Residential burglary	%	
First British Crime Survey	70	(Hough and Mayhew 1983: 12)
Second British Crime Survey	70	(Hough and Mayhew 1985: 61)
Third British Crime Survey	65	(Mayhew *et al.* 1989: 70)
Oxford and Salford	76	and 80% (McCabe and Sutcliffe 1978: 55–6)
London survey (burglary/ theft from dwelling)	40	(Sparks *et al.* 1977: 157)
Islington Crime Survey	52	(MacLean *et al.* 1986: Table 3.5)

Violent crime		%
Second British Crime Survey	Theft from person/robbery	25
	Sexual offences	133
(Hough and Mayhew 1985: 61)		
Third British Crime Survey	Robbery	38
	Wounding	49
	Sexual offences	77
(Mayhew *et al.* 1989: 70)		
Oxford and Salford	Assault	50 and 29
(McCabe and Sutcliffe 1978: 55–6)		
London survey	Assault/theft person	13
(Sparks *et al.* 1977: 157)		
Islington Crime Survey	Robbery	76
	All assault	16.7
	All sexual offences	43
(MacLean *et al.* 1986: Table 3)		

Thus in general it can be said that there are significant problems of reporting and recording especially with regard to violent crime. How far do victimization surveys provide a satisfactory alternative? Problems exist with such surveys, some of which are specific to victimization surveys.

Victimization surveys

A first problem which may be identified is that of coverage in that usually only some crimes rather than all crimes are included in such surveys. The kinds of crimes normally included are those with a clearly identifiable victim, such as street crime.

The 'halo' effect may also be evident in the sense that the interviewer may be seen as 'demanding' details of victimizations. Some surveys such as the second British Crime Survey use a two-stage interview procedure, initially checking off crime incidents and later recording their details. This can be seen as taking off the pressure on the respondents. The third problem of victimization surveys are memory effects. Are crimes memorable incidents? Skogan (1986: 87) quoting Biderman noted that 'respondents have to do a great deal of thinking and slow reflection before they can remember even fairly serious crimes of which they were victims some time ago'. A second aspect of memory effects is the possibility of lying by respondents about, for example, non-stranger assaults and a third aspect is the telescoping of victimization incidents which may involve bringing forward 'old' victimizations into the survey's reference period. The results of a study by Schneider which compared survey and police data are quoted by Skogan (1986: 93). The study found that 'on the average, matched incidents were pulled forward within the period by 2.2 months. Forty-nine per cent of all incidents were placed in the wrong months by their victims'.

We should recognize, as does Young (1988a: 168), that 'victimization statistics like police figures, have dark figures'. This applies particularly for sexual assault and domestic crime. In a review of the response rates in eleven victimization surveys Young (1988a: 168) shows a variation of 57–86 per cent. As he then says 'such a large unknown population could easily skew every finding that we victimologists present' (Young 1988a: 169).

A further problem with victimization surveys is that such surveys have failed to provide any detailed information on multiple victimization. This is important with regard to left realism (as will be explained later). Three problems are apparent for surveys wishing to record multiple victimizations. The first is that surveys conceptualize crimes as discrete incidents rather than as a process so that the questions are designed to elicit information about such incidents. The second problem is that as Genn (1988: 91) says 'victims' experiences of crime may be so common, or individual incidents so similar, that respondents to survey questionnaires cannot recall dates or details of the relevant events to be recorded'. Finally in order not to inflate gross victimization rates arbitrary upper limits are placed upon series offences. Thus the British Crime Survey imposes an upper limit of five (Hough 1986).

A final problem of victimization surveys is the different perceptions that respondents may bring to the same question. Thus Hough (1986: 119) has for instance noted that in the first British Crime Survey 'those with more education were most likely to report incidents of violent victimisation'. He also states that 'similar findings have not emerged in the British crime survey for burglary, car theft and other household crimes'. As an explanation of higher reporting rates of violent incidents by those with more education Hough suggests that those with more education may apply lower threshold of seriousness when it comes to defining violent incidents.

PROPOSITION TWO: THE WORKING CLASS SUFFER DISPROPORTIONATELY FROM ALL KINDS OF CRIMINAL ACTIVITY

Lea and Young (1984: 264) state that 'In contrast to the beliefs of left realists, working-class crime really is a problem for the working-class . . . left realism notes that the working-class is a victim of crime from all directions; that one sort of crime tends to compound another'. How far is this statement true? The results of a variety of victimization surveys give us some indication of the likely victims of particular crimes.

Residential burglary

With regard to residential burglary the 1980 General Household Survey (OPCS 1982: 80) shows a relatively clear association between status of the household head and burglary rates. In general the lower social categories had higher burglary rates than the higher categories, alhough the gradient of rates is not a perfect one. By contrast the first British Crime Survey (Hough and Mayhew 1983: 19) found no clear relationship between burglary rates and a household's social class. Gottfredson (1984), who extended the analysis of the first British Crime Survey, surprisingly includes no social class analysis of the victims of household crime.[1] This is the case even though demographic attributes are included in the analysis. Gottfredson does, however, point out that inner city residents, households who live in flats and council tenants are more at risk from household crime.

The Islington Crime Survey (MacLean *et al.* 1986), conducted in a generally but not entirely deprived area, found that within this 'inner city' area it was the households with the highest earners which had the highest burglary rates.

Spatial studies while being subject to the ecological fallacy point to the vulnerability of lower-income households to residential burglary. The Merseyside Crime Survey (1984) surveyed residents of five areas in Merseyside. Two of these areas were affluent areas, two were inner city areas and one was a local authority housing area (LAHA). The respective burglary rates for a twelve-month period were 2 per cent and 1 per cent (affluent areas), 13 per cent and 8 per cent (inner city areas) and 6 per cent (LAHA area). In Herbert's (1982: 58) study of residential crime (residential burglary and thefts from dwellings) in west Swansea the vulnerable areas were in the heterogeneous inner city or some large public sector estates, not the 'prime "target" areas or the wealthier suburbs'. Evans and Oulds (1984) in their analysis of residential burglary rates in Newcastle under Lyme found eight enumeration districts with high rates on the overall burglary measure; each of these areas was of low socio-economic status. Only with regard to high-value burglaries (£100 and over stolen) did middle-class areas record high rates. In that case five of twenty-three enumeration districts with high rates were middle class. In Maguire's (1982) analysis of burglary in Reading:

> The greatest concentration [of burglaries recorded by the police] occurred in residential areas close to the town centre and in the large council housing estates to the south. The middle class areas to the north escaped comparatively lightly. ... in the centre of town Reading burglars selected a disproportionate number of high-value properties.
>
> (Maguire 1982: 37)

The second British Crime Survey (Hough and Mayhew 1985: 37) indicates that the high-risk areas for residential burglary in terms of the ACORN classification are high-status non-family areas, multiracial areas and the poorest council estates.

Generally poorer households are more at risk from residential burglary, although within poor areas, higher value properties may be more at risk from residential burglary.

Violent crime

Violent crime is a particularly heterogeneous crime category so the discussion will focus upon it both as a general category and then upon more specific forms of violent crime.

Gottfredson (1984: 8) indicates that for personal crime[2] the risk is greater for unemployed people rather than employed people and that the risk for those in full-time education is much higher than for either of those two groups. Also that the risks are higher for residents of the inner cities than in other parts of metropolitan areas, small cities or rural areas. Again no direct analysis of social class is included. Similarly Walmsley (1986: 38) suggests that the risk of personal violence[3] is higher for the unemployed. In the Islington Crime Survey (MacLean et al. 1986: 3.29) the highest assault rate was in the £8,000–11,999 income bracket, the next-to-highest category in the survey. Hough (1986) and Hough and Sheehy (1986) provide some direct analysis of the relationship between being the victim of a violent crime and social class. In an analysis of the first British Crime Survey the percentage of male respondents who were victims of violent offences showed little difference between non-manual, manual and unemployed respondents, although students and others had a 9 per cent higher rate of victimization (Hough 1986: 120). The lack of a difference between non-manual and manual respondents may owe something to the former's greater likelihood of reporting violent incidents to the interviewer, for the reasons noted earlier. In an analysis of the results of the first and second British Crime Surveys, Hough and Sheehy (1986: 23) note that while manual male and female respondents aged 16–30 are more likely to be the victim of non-stranger assaults, there was little difference in the rates of stranger assaults, and mugging for manual and non-manual respondents. Davidson (1989: 64), using the ACORN classificaton, identifies the high-risk neighbourhoods for assault as poor quality older terraced housing, high-status non-family areas, multiracial areas and the poorest council estates.

Evidence for the social classification of the victims of violent crime is not comprehensive but there are some indications that poorer people are more at risk. This is shown with the association between violent crime and unemployed persons, with inner city residents and in the victimization rates for non-stranger assaults.

For the crime of robbery information from the British Crime Surveys with regard to the social class of victims is sparse. The main characteristics of robbery victims which are given are the age and sex of the victims (Hough and Mayhew 1983: 17; Walmsley 1986: 35). These surveys also suggest that inner city residents are more at risk from this offence (Hough and Mayhew 1983: 17; Walmsley 1986: 38). The second British Crime Survey (Hough and Mayhew 1985: 73) specify three types of areas as having high rates of robbery and snatch theft, these being those described earlier as having high rates of burglary. Davidson (1981) suggests that

> Typically muggings are perpetrated by young males, on older males and there is a racial element in the attack. Since, as with many young offenders, muggers do not mug far from home, many of their victims come from the same sort of neighbourhood. Thus muggings tend to be concentrated in poorer inner-city districts.
>
> (Davidson 1981: 35)

In the Islington Crime Survey (MacLean *et al.* 1986: 3.21) the highest income category had the highest rate of theft from the person (including robbery) and employed persons a higher rate than those unemployed. Inner city residents, who are mainly but not all poor, are most at risk from robbery. As with burglary, within inner city areas richer residents are more frequently victimized.

For the crime of sexual assault the Islington Crime Survey (MacLean *et al.* 1986: 3.26) indicates that the second highest income category in their survey (£8,000–11,999) has the highest rates of sexual assault and that employed rather than unemployed people have a higher rate. No information on the social class of the victims of sexual assault is provided by the British Crime Surveys.

Although the evidence is not as specific as it might be there are at least some indications of a relationship between violent crime and robbery and low socio-economic status.

Motor vehicle theft and vandalism

For the crime of motor vehicle theft Gottfredson (1984: 28) indicates that residents of inner cities, flats and maisonettes and council tenants are more at risk from this particular crime. Again these surrogate measures may indicate a relationship between this particular crime and low socio-economic status.

For the crime of vandalism Gottfredson (1984: 22) indicates that, in terms of tenure, those in owner-occupied housing and 'other' tenure categories are most at risk. In this case the difference in rates between inner cities and other urban areas does not occur. In the Islington Crime Survey (MacLean *et al.* 1986: 3.17) the highest rate of vandalism is experienced by the highest income earners.

Subgroup analysis

One can also take this analysis of the characteristics of victims further. The rationale for doing so is provided by statements from the left realist writers. Thus Young (1988a: 171) states that 'a realist criminology must start from the actual sub-groups in which people live their lives, rather than from broad categories which conceal wide variations within them'. Similarly Lea and Young (1984: 27, 38) 'crime is a serious threat only to those in small marginal localities and in certain social categories' and that with regard to the localization of crime 'such localisation can involve quite extreme foci'. These statements suggest that the analysis of victims' characteristics including location should be at a detailed level and that the importance of multiple victimization should be acknowledged.

The Islington Crime Survey (Jones *et al.* 1986: 49) does analyse victimization within small subgroups. Thus the incidence of burglary is shown to be high at both the lower and higher ends of the income bands used in the survey while when burglary rates are examined by age, race and gender the highest rates are those experienced by Asian men aged 16–24 years and black women aged 45 years or more (Jones *et al.* 1986: 49–50). Theft from the person is highest at the highest income category used and for black, Asian and non-white women aged 25–44, 16–24 and 25–44 respectively (Jones *et al.* 1986: 59–60). A third example is assault, which is highest for the second highest income group of the scale used in the survey and is highest for black and white men and women aged 16–24 years and is also high for black women aged 25–44 years (Jones

46

et al. 1986: 67). There is therefore considerable variation between specific subgroups of criminal victimization rates.

An independent race effect

In addition to class and subdivisions of classes, race has been shown to be an important independent factor affecting some types of victimization. Thus Mayhew *et al.* (1989: 45) say that the results of the third British Crime Survey show that 'factors to do with area of residence and other demographics, however, did not explain higher risks of vandalism among Asians: ethnicity was an important factor. This was also true in relation to contact thefts against Asians'. However, 'In contrast, differential risks between Afro-Caribbeans and whites were more likely to be accounted for by differences other than ethnicity'. Thus an independent race effect was found only for Asians and only for two specific crimes.

Multiple victimization

The importance of multiple victimization is revealed by a number of studies. Thus the Islington Crime Survey (Jones *et al.* 1986: 47) indicates that 'the best predictor of whether one's household will be burgled or not is whether or not it has been burgled already'. This is because the 'amateur burglars tend to break into the same places over and over again because they found that they were not caught there before, or that the household, by experience, was an easy target'. Again in relation to burglary, in the specific context of a large local authority estate, Forrester *et al.* (1988: 9) say that 'on Kirkholt in 1986, the probability of a home being burgled for a second time was four times the expected rate' calculated in a random manner. They also provide data upon the high degree of multiple victimization for burglary and theft in a dwelling revealed in data drawn from the first and second British Crime Surveys (Forrester *et al.* 1988: 9).

On a broader level for several crimes the importance of multiple victimization is shown at least for inner city areas by both the Islington Crime Survey (Jones *et al.* 1986: 84) and the Hammersmith and Fulham crime and policing survey (Painter *et al.* 1989: 11). Barr and Pease (1990: 40) calculate the rates of multiple victimization for different crimes using data from the 1982 and 1984 British Crime Surveys. For a second victimization the difference between observed and expected frequencies is over five to one for domestic burglary with loss, theft from the person and violence. For a third victimization equivalent ratios are over ten to one for domestic burglary with loss, violence and theft from a motor vehicle.

PROPOSITION THREE: A HIGH INCIDENCE OF FEAR OF CRIME DIMINISHES THE QUALITY OF COMMUNITY LIFE FOR WORKING-CLASS GROUPS

Lea and Young (1984: 264) state that 'fear of street crime helps the disintegration of working-class communities'.

Levels of fear of street crime

The levels of fear about street crime may be seen from the results of recent victimization surveys. The first and second British Crime Survey included the question, 'How safe do you feel walking alone after dark?' In 1981 34 per cent of the respondents said that they felt either 'a bit' or 'very' unsafe; the corresponding figure in 1983 was 31 per cent. For men the respective figures were 14 and 13 per cent; for women they were 51 and 48 per cent. For worry about particular crimes, the rank order of levels of anxiety was rape, 30 per cent; sexual harrassment, 28 per cent; burglary, 23 per cent, vandalism, 22 per cent; mugging, 20 per cent; stranger attacks, 18 per cent; stranger insults, 10 per cent (Hough and Mayhew 1985: 71–2). The Islington Crime Survey (Maclean et al. 1986: 2.3) indicated levels of worry about particular crimes. The percentages worrying 'quite a bit' or 'a lot' about individual crimes was as follows: mugged and robbed, 46 per cent; being raped, 45.5 per cent (women only), sexually molested, 44.5 per cent (women only), attacked by strangers, 38.7 per cent. Finally the Merseyside Crime Survey (Kinsey 1984) gives levels of worry about street crime at 45 per cent of the respondents.

These surveys suggest between approximately one-third and one-half of the population are worried about the possibility of being the victim of street crime.

The impact of fear of street crime

Fear of street crime may lead to avoidance behaviour on the part of those most fearful of street crime. The second British Crime Survey

(Hough and Mayhew 1985: 40) reports that 2 per cent of the whole sample never went out after dark in whole or in part because of fear of crime. Those in high crime areas were over-represented (4 per cent) and among elderly women in these areas the figure was 18 per cent. The Islington Crime Survey (MacLean *et al.* 1986: 5, 12) reports that 23.9 per cent of the respondents often avoid going out after dark as a precaution against crime and that 43.6 per cent of the respondents always take this precaution. This gives a total of 67.5 per cent of the respondents (53.6 per cent of whom are women) who often or always avoid going out after dark as a precaution against crime. The Merseyside Crime Survey (Kinsey 1984) gave a figure of 24 per cent of the respondents to a similar question.

This kind of avoidance behaviour, particularly in inner city areas where the figures are very much higher, is obviously not conducive to community integration.

However, there are opposing views of the impact of fear of crime upon community integration. Smith (1983: 236) states that 'crime can be a source of group cohesion when the very violation of institutionalised norms prompts collective action on the part of the majority which both signifies and reinforces social solidarity'. She is here, as she says, reflecting the views of Durkheim, Mead and Simmel. Smith also gives the opposing view of Conklin when she says that 'as a consequence of the suspicion and mistrust generated by fear, crime is instrumental in bringing about social disintegration and disorganisation'. Her empirical evidence from Birmingham is sharply divided on this question. In response to the question that 'people know one another and help one another', 41.2 per cent of the respondents said that this always occurred and 34.5 per cent said that this never occurred. In response to the question 'people keep themselves to themselves', 33.3 per cent of the respondents said that this always happened and 29.4 per cent said that it never happened. As Smith (1983: 236) says, 'The evidence in Birmingham, however, is that both conditions affect the community simultaneously'. Two further empirical studies may be cited concerning the effect of fear of crime upon community integration. The first by Conklin (1971) was concerned with a sample of respondents in a high crime rate urban area and a low crime rate suburban area. It showed lower levels of trust and dependability in the high crime rate area (Conklin 1971: 381). The second study of Hartnagel (1979: 189), which is again a North American study, found no relationship between fear of crime on the one hand and indicators of neighbourhood cohesion and social activity on the other.

49

There is therefore little firm evidence of the impact of fear of crime upon community integration. Fear of crime undermines community integration, especially in places like Islington, given the high levels of behavioural modification in response to the fear of crime. Further empirical studies are needed to examine this relationship.

The fear/risk equation

Later work by the left realist Young (1988a) has questioned some of the basic findings of fear of crime studies.

Thus fear of crime studies often state that fear is irrational in that those most likely to be victimized are least afraid and vice versa (see, for instance, Maxfield 1984: 38; Smith 1987: 5).

Young (1988a) has pointed out that these statements are subject to heavy qualification. Thus with regard to women and basing his comments on the Islington Crime Survey Young (1988a) says that 'Our survey suggests that their [women's] fears are perfectly rational. For women are, in fact, more likely to be victims of crime than men'. This applies to data regarding not only sexual assault, but also vandalism, theft from the person and assault. Similarly with regard to elderly people Young says that the data from the Islington Crime Survey show that when someone over 45 is assaulted the attack is more likely to have a greater effect on their lives in terms of level of injury and amount of time off work (Young 1988a: 170). Young also questions the validity of the fear/risk equation, which he feels is flawed because it assumes 'that there is an objective crime rate irrespective of the subjective assessment of various sub-groups' (Young 1988a: 173). He goes on to say that 'the "same" punch can mean totally different things in different circumstances' and that 'Violence, like all forms of crime, is a social relationship. It is rarely random: it inevitably involves particular social meanings and occurs in particular hierarchies of power' (Young 1988a: 174). Any crude fear/risk equation which uses victimization rates ignores the particular circumstances of criminal acts.

PROPOSITION FOUR: THE CAUSES OF CRIME ARE MACRO AND MICRO IN NATURE

Introduction

In relation to the causes of crime Lea and Young (1984: 265) state 'Its cause is seeing through the deception and inequality of the world; its direction is towards that of selfishness. Its cause is righteous, its direction individualistic'.

Lea and Young (1984: 95) thus feel that crime is a solution for some individuals to the levels of relative deprivation inherent in the capitalist system. It is a solution to the contradiction between 'the values of an equal or meritocratic society which capitalism inculcates into people' and 'the actual material inequalities in the world'. The key words in Lea and Young's approach to the causes of crime are subculture, relative deprivation and marginalization. The explanation, therefore, includes macro and micro aspects.

The social geographer would comment that Lea and Young's statements on the causes of crime are not area specific and that offender rates in socially similar areas can be different due to specific geographical influences such as the labelling of the area by its external caretakers, the housing allocation policies of the local authority which may reflect such labelling and the detailed variation in attitudes between socially similar areas. These are significant micro aspects which affect the offender rate of an area. The technique of matched pairs, which has been used by Herbert, Evans and Bottoms, allows the offending experiences of socially similar areas to be examined.

Empirical aspects

The Cardiff study of juvenile delinquency (Evans 1980) does offer some limited support for the subculture theory. In that study attitudes to certain minor delinquent acts were different between the socially matched delinquent and non-delinquent areas. There were also differences between the areas in the means of punishing misbehaviour and in the amount of control of children's behaviour. In addition the attitudes to and attainment in education was different between socially similar areas.

Herbert (1982: 93, 99) using the results of the same study showed that the residents' perceptions of delinquency areas fitted their classification through official data sources. Herbert (1982: 98) considers that Ely, a local authority housing area with a high delinquency rate, is stigmatized as a problem estate and that housing policy and tenant selection contribute to this stigmatization. The same is not true of Mynachdy which occupies a similar social position to Ely.

Bottoms and Xanthos (1981) also use the technique of matched pairs to investigate the offender rates of samples of areas. In the pair of housing areas the difference in offender rates was attributed to the indirect effects of the local authority allocation system, the strong importance of reputation and its effects, the development of a subcultural value system, socialization into this value system and a differential influence of the two schools on the two estates. The explanation for the difference in offender rates for the flat areas was said to be related to the concept of social instability.

PROPOSITION FIVE: CONVENTIONAL TARGET-HARDENING CRIME PREVENTION MEASURES DO NOT DIRECTLY ADDRESS THE CRIME PROBLEM BUT POLICIES WHICH CREATE A STRONGER SENSE OF COMMUNITY ARE EFFECTIVE

Lea and Young (1984: 267) state that 'Environmental and public precautions against crime are always dismissed by left idealists and reformers as not relating to the heart of the problem. They are distractions from real concerns and furthermore, because they do not get at the causes of crime, are largely irrelevant'. Lea and Young (1984: 267) also state that 'the organisation of communities in an attempt to pre-empt crime is of the utmost importance'.

The effectiveness of neighbourhood watch

Herbert and Harries (1986: 284) suggest a threefold typology of area-based crime prevention policies, namely those concerned with tactical policing, physical measures and social awareness. This typology may be used as a framework to assess the relative effectiveness of different area-based crime prevention measures. The third category, namely that of social awareness measures, includes neighbourhood watch (NW), which has enjoyed a dramatic growth in Britain since the mid-1980s. Husain (1988: 1) says that 'NW has spread through Britain with astonishing rapidity. Virtually unknown four years ago, by 1987 there were 42,000 schemes in operation (*Hansard* 1987). It seems likely that more than 2.5 million households now live within a NW area'. The effect of such schemes upon crime rates is at present unclear. One initial comment that can be made is that the schemes are more common in low-risk areas and less frequent in high-risk areas especially the poorest council estates and multiracial areas (Husain 1988: 11). One of the more sophisticated tests of such schemes to date is by Bennett (1988). His study which was conducted in London identified two areas which were about to implement NW and compared these with an adjacent displacement area and a more distant non-NW control area. Data through questionnaire survey were collected from the four areas before the launch of NW schemes and again one year following the launch. After controlling for sample differences the general finding was that the

> victimisation rates increased in the NW areas. In both areas the increase was statistically significant. Victimisation rates in the displacement area remained static from the 'before' to the 'after' surveys and decreased in the control area. There was no evidence, therefore, from these results of a NW effect on victimisation.

The 1988 British Crime Survey (Mayhew *et al.* 1989: 57, 58) provides three tests of the effectiveness of neighbourhood watch in reducing burglary rates. Two of the three tests showed no reduction of burglary rates for NW members. The two tests which show no improvement are the cross-sectional comparisons of the burglary rates for NW members and non-members; one of these comparisons being a matched comparison. Longitudinal data, however, show that 'Members ... show a larger drop in risks (p. <0.05), suggesting that joining NW made them less vulnerable than previously'. An analysis by Forrester *et al.* (1988) upon the effect of crime prevention measures implemented in a local authority estate in Rochdale did show a decrease in overall levels of burglary and of multiple victimization. The measures implemented included 'cocoon' neighbourhood watch in addition to the uprating of household security, property postcoding and removal of gas and electric cash pre-payment meters. Although in total the measures were effective in the period immediately following implementation because of the mix of measures used it is not possible to specify the impact of each measure. What can be said is that the neighbourhood watch was implemented on a smaller more 'organic' scale and that this may be a more effective scale at which to implement these schemes.

Further rather negative results about the effectiveness of NW in reducing burglary rates have been found in USA. Thus in a matched comparison of two middle-class suburbs, one of which had adopted the Alert neighbours scheme and one of which hadn't, it was found that the:

Alert neighbours ... seemed to have had little impact on the incidence of property crimes in the target area. This area actually reported more burglaries during the operation of the project than did the control area, though levels of offences were really very low in both areas.

(Herbert and Harries 1986: 290)

One possible explanation for the negative findings concerning the effectiveness of neighbourhood watch is related to tolerance levels of behaviour.

As Young (1988b: 179) says 'the real number of crimes is a product of behaviour and tolerance levels' and 'a successful [crime prevention] campaign may easily create a decline in tolerance to these offences, and thus a rise in the number of offences occurring'. Young was here referring to crimes such as vandalism and minor violence but the argument may also be extended to include attempted and low-value burglaries.

The effectiveness of physical measures of crime prevention

The evidence for the effectiveness of physical measures and their impact upon crime rates is similarly mixed. Allat (1984) studied the effect of target hardening upon a difficult-to-let estate. Although there was no decline in burglary upon the target estate a trend analysis of the monthly data revealed a pronounced levelling off for just over a year following the start of security improvements. This contrasted with the encompassing police subdivision and the control area where burglary rose substantially. There was also some displacement of burglary to an area of private housing and an adjacent council estate. In addition there was also displacement to other crimes within the target estate; other crimes, such as theft of vehicle and burglary of other premises. The evidence for the effect of property marking upon crime rates provided by Laycock (1985) are encouraging although the results have to be seen within a specific context. The property marking project was set up in a relatively isolated area in South Wales consisting of three distinct villages, one of which, Craig-y-Rhacca had the highest burglary rate. Burglaries were reduced by 40 per cent in the area in the twelve-month period after the launch. There were statistically significant reductions in the victimization rate in Machen/Trethomas, Craig-y-Rhacca and in the valley as a whole following the launch of the scheme. There was also a reduction in the number of burglary incidents for those participating in the scheme while there was no reduction for non-participants. However, one of the final comments is very important in assessing the value of property marking in reducing burglary.

> This initiative was set up as a demonstration project in a carefully controlled and selected rural area. It was comparatively isolated and as such it was possible to ensure that potential burglars were as informed about the project as were the rest of the community. This is a highly significant factor and is unlikely to be possible to the same extent elsewhere.
>
> (Laycock 1985: 14)

It is interesting to note also that the reduction in burglary for Craig-y-Rhacca involved not only markable goods but also other goods which were not marked. As Laycock says (1985: 12) this may be because 'the message to potential burglars that the risk of breaking into this home is greater than that associated with another "unmarked" home'. A further appraisal of the effect of physical measures upon crime rates is provided by Painter (1988). Essentially the project was a 'before' and 'after' study of the impact of lighting on crime. People were asked especially about their experiences of crime and harassment in one road, six weeks before and six weeks after lighting had been installed. A total of twenty-one incidents of assault, auto-crime and threats were reduced to three (Painter 1988: 71). Painter (1988: 75) also provides a robust defence of her scale of study when she says

> The strength of our evidence does not lie in weighty, random samples which are strong on statistical significance and randomisation, and weak on control. Tackling small numbers of crimes on a local basis presents a far more effective approach to crime prevention.

A final comment upon the effect of target hardening is provided by Winchester and Jackson (1982). In this analysis of residential burglary in suburban Kent 'environmental risk' was the most important factor in distinguishing between a sample of 491 non-victim and 434 victim households. Thus the authors say that

> The most important factor in making the discrimination between victim households and households who had not been

burgled proved to be environmental risk, followed by occu-
pancy rates and reward in this order. Relative security levels
did not contribute to this discrimination.

(Winchester and Jackson 1982: 22).

Specifically with regard to security levels Winchester and Jackson
(1982: 21) say 'From this police evidence alone, it is clear then that
burglars can and will break into houses with good levels of security
protection. Even a burglar alarm may not guarantee protection
against burglary'. This comment runs counter to the comments
made by burglars in Maguire's (1982) and Bennett and Wright's
(1984) behavioural analysis.

The effectiveness of policing strategies

With regard to policing this was used as part of a three-pronged
approach to crime reduction in the Asylum Hill area of Hartford,
Connecticut (Fowler and Mangione 1986). The three approaches
used were changes to the physical environment, changes to policing
and the formation of a residents' organization. Initially the crime
rate was reduced but then it was found that burglary rates in the
area increased while they declined in the rest of Hartford and that
the robbery/purse-snatching rate rose to a level that would have
been predicted without the programme. Two comments suggest
that changes in policing may have brought about these trends. First,
Fowler and Mangione say that 'There was a significant erosion of
the police team components of the program between 1977 and 1979'
(Fowler and Mangione 1986: 95, 105). Studies conducted in Newark
and Flint, Michigan (Rosenbaum 1986) have produced contrasting
evidence for the effectiveness of increasing policing in order to
reduce crime. While in Newark it was found that the addition of
intensive foot patrol coverage appeared to have no significant effect
on victimization, in Flint the neighbourhood foot patrol programme had the impact of markedly reducing crime rates in a three-
year evaluation period.

CONCLUSION

Left realism is a criminological perspective which can assist in the
spatial study of crime. The victimization surveys even with their
limitations provide a ready source of information about various

aspects of crime. The victims of crime are overwhelmingly of low social status. Subgroup analysis, the existence of an independent race effect for specific crimes and the importance of multiple victimization are further caveats produced by victimization studies. Left realists have modified the irrationality statements concerning the fear of crime and risk of crime equation. The macro and micro distinction which left realists adopt with regard to the causes of crime enable the matched pairs technique to be used to add micro-scale spatial detail to causation theories. Finally researchers should examine the most effective ways in which communities can organize in order to pre-empt crime. In these ways left realism can provide a body of theory and practice which can assist in the spatial study of crime.

NOTES

1 Gottfredson (1984: 3) includes vandalism, burglary in a dwelling, bicycle theft, theft in a dwelling and other household thefts in the term 'household crime'.
2 Gottfredson (1984: 2) includes common assault, theft from the person, wounding, robbery, sex offences and other personal theft in the term 'personal crime'.
3 Walmsley (1986: 1) includes the crimes of violence against the person, robbery and sexual assault in the term 'personal violence'.

REFERENCES

Allat, P. (1984) 'Residential security: containment and displacement of burglary', *Howard Journal* 23 (2): 99–116.
Barr, R. and Pease, K. (1990) 'Crime displacement and placement', in M. Tonry and N. Morris (eds) *Crime and Justice*, vol. 12, Chicago: University of Chicago Press.
Bennett, T. (1988) 'An assessment of the design, implementation and effectiveness of neighbourhood watch in London', *Howard Journal* 27 (4): 241–55.
Bennett, T. and Wright, R. (1984) *Burglars on Burglary*, Aldershot: Gower.
Bottomley, A.J. and Coleman, C.A. (1981) *Understanding Crime Rates*, Aldershot: Gower.
Bottoms, A.E. and Xanthos, P. (1981) 'Housing policy and crime in the British public sector', in P.J. Brantingham and P.L. Brantingham (eds) *Environmental Criminology*, Beverly Hills, Calif.: Sage.
Bottoms, A.E., Mawby, R. and Walker, M.A. (1987) 'A localised crime survey in contrasting areas of a city', *British Journal of Criminology* 27: 125–54.
Conklin, J.E. (1971) 'Dimensions of community response to the crime problem', *Social Problems* 18: 373–85.

Davidson, R.N. (1981) *Crime and Environment*, London: Croom Helm.
—— (1989) 'Micro-environments of violence', in D.J. Evans and D.T. Herbert (eds) *The Geography of Crime*, London: Routledge.
Evans, D.J. (1980) *Geographical Perspectives on Juvenile Delinquency*, Farnborough: Gower.
Evans, D.J. and Oulds, G. (1984) 'Geographical aspects of the incidence of residential burglary in Newcastle under Lyme, UK', *TESG* 75: 344–455.
Forrester, D., Chatterton, M. and Pease, K. (1988) *The Kirkholt Burglary Prevention Project, Rochdale*, Crime Prevention Unit Paper 13, London: Home Office.
Fowler, F.J. and Mangione, T.W. (1986) 'A three-pronged effort to reduce crime and fear of crime: The Hartford Experiment', in D.P. Rosenbaum (ed.) *Community Crime Prevention: Does it Work?*, Beverly Hills, Calif.: Sage.
Genn, H. (1988) 'Multiple victimisation', in M. Maguire and J. Pointing (eds) *Victims of Crime: A New Deal*, Milton Keynes: Open University Press.
Gottfredson, M.R. (1984) *Victim of Crime: The Dimensions of Risk*, Home Office Research Study 81, London: HMSO.
Hartnagel, T.F. (1979) 'The perception and fear of crime: implications for neighbourhood cohesion, social activity and community effect', *Social Forces* 58: 176–93.
Herbert, D.T. (1982) *The Geography of Urban Crime*, Harlow: Longman.
Herbert, D.T. and Harries, K.D. (1986) 'Area-based policies for crime prevention', *Applied Geography* 6: 281–95.
Hough, M. (1986) 'Victims of violent crime, findings from the British crime survey', in E.A. Fattah (ed.) *From Crime Policy to Victim Policy*, Basingstoke: Macmillan.
Hough, M. and Mayhew, P. (1983) *The British Crime Survey: First Report*, Home Office Research Study 76, London: HMSO.
—— (1985) *Taking Account of Crime: Key Findings from the Second British Crime Survey*, Home Office Research Study 85, London: HMSO.
Hough, M. and Sheehy, K. (1986) 'Incidents of violence: findings from the British crime survey', *Home Office Research and Planning Unit Bulletin*, 20: 22–6.
Husain, S. (1988) *Neighbourhood Watch in England and Wales: A Locational Analysis*, Crime Prevention Unit Paper 12, London: Home Office.
Jones, T., MacLean, B. and Young, J. (1986) *The Islington Crime Survey*, Aldershot: Gower.
Kinsey, R. (1984) *Merseyside Crime Survey: First Report, November 1984*, presented by R. Kinsey, Centre for Criminology, University of Edinburgh.
Laycock, G. (1985) *Property Marking: A Deterrent to Domestic Burglary?*, Crime Prevention Unit Paper 3, London: Home Office.
Lea, J. and Young, J. (1984) *What is to be Done about Law and Order?*, Harmondsworth: Penguin.
McCabe, S. and Sutcliffe, F. (1978) *Defining a Crime: A Study of Police Decisions*, Occasional Paper 9, Centre for Criminological Research, Oxford University.

Maclean, B.D., Jones, T. and Young, J. (1986) *Preliminary Report of the Islington Crime Survey*, Centre for Criminology and Police Studies, Middlesex Polytechnic.

Maguire, M.(1982) *Burglary in a Dwelling*, London: Heinemann.

Maxfield, M.G. (1984) *Fear of Crime in England and Wales*, Home Office Research and Planning Unit Report 78, London: HMSO.

Mayhew, P., Elliott, D. and Dowds, L. (1989) *The 1988 British Crime Survey*, Home Office Research Study 111, London: HMSO.

OPCS (1982) *General Household Survey 1980*, London: HMSO.

Painter, K. (1988) *Lighting and Crime Prevention: The Edmonton Project*, Centre for Criminology and Police Studies, Middlesex Polytechnic.

Painter, K., Lea, J., Woodhouse, T. and Young, J. (1989) *Hammersmith and Fulham Crime and Policing Survey Final Report*, Centre for Criminology and Police Studies, Middlesex Polytechnic.

Rosenbaum, D.P. (1986) *Community Crime Prevention: Does it Work?*, Beverly Hills, Calif.: Sage.

Skogan, W.G. (1986) 'Methodological issues in the study of victimisation', in A.E. Fattah (ed.) *From Crime Policy to Victim Policy*, Basingstoke: Macmillan.

Smith, S.J. (1982) 'Race and reactions to crime', *New Community* 10: 233–42.

—— (1983) 'Public policy and the effects of crime in the inner cities: a British example', *Urban Studies* 20: 229–39.

—— (1987) 'Fear of crime: beyond a geography of deviance', *Progress in Human Geography* 11 (1): 1–23.

Sparks, R.F., Genn, H.G. and Dodd, D.J. (1977) *Surveying Victims*, Chichester: Wiley.

Tuck, M. and Southgate, P. (1981) *Ethnic Minorities, Crime and Policing*, Home Office Research Study 70, London: HMSO.

Walmsley, R. (1986) *Personal Violence*, Home Office Research Study 89, London: HMSO.

Winchester, S. and Jackson, H. (1982) *Residential Burglary: The Limits of Prevention*, Home Office Research Study 74, London: HMSO.

Young, J. (1987) 'The tasks facing a realist criminology', *Contemporary Crimes* 11: 337–56.

—— (1988a) 'Risk of crime and fear of crime: a realist critique of survey-based assumptions', in M. Maguire and J. Pointing (eds) *Victims of Crime: A New Deal*, Milton Keynes: Open University Press.

—— (1988b) 'Radical criminology in Britain: the emergence of a competing paradigm', *British Journal of Criminology* 28 (2): 159–83.

3

LOCAL AREA PROFILES OF CRIME
Neighbourhood crime patterns in context
Norman Davidson and Trevor Locke

The widespread introduction of computer-based crime logging systems in recent years has meant that information about patterns of crime is now often available down to police beat level. Parallel to this is a growing recognition at all levels from government down that the crime problem will not be solved without the consent and active participation of the public. Indeed wide-scale participation can be achieved only at the local community level where it has a firm foundation in real concerns – the basis, for example, of the success of neighbourhood watch (NACRO 1988). In this chapter we wish to show how the conjunction of crime information and crime initiatives at the community level highlights the need for and roles of local crime profiles.

THE NEED FOR LOCAL CRIME PROFILES

Crime is a very unevenly distributed phenomenon. Whatever geographical scale is used, disparities emerge consistently, both in overall levels of crime and in specific categories of offence. However, these disparities show a tendency to grow as the spatial resolution is increased. Neighbourhoods within a city are more disparate, on the whole, than cities within a region or regions within a country. Thus in England and Wales, the police force areas with the highest crime rates have about three times the rate of those with the lowest (Home Office 1989). Within a single city such a ratio may exceed ten (Mawby 1979; Davidson 1981).

Investigation of these disparities has provided the basis for a long-standing theme in criminology: the ecological perspective, which is currently undergoing something of a revival (Reiss 1986; Locke

60

1990b). The focus of renewed interest is in the concept of 'community careers' in crime (Bottoms and Wiles 1986), which postulates that neighbourhoods have an effect on criminal behaviour quite separate from and independent of those of family, status, peer-group or criminal experience. Such effects are provided by mechanisms that operate both directly on crime, for example local opportunity structures, and indirectly, for example housing management or facility location processes. As with criminal careers themselves, the dynamic element of community crime careers is emphasized, not least in the way that communities respond to crime or the threat of crime. It is this premise of the importance of understanding local crime rates and their perceptions which underwrites the attempt in this chapter to provide a model for unravelling the significance of levels of recorded crime in a particular community.

The need for local profiling has also emerged from two different, more pragmatic directions. One relates to greater emphasis on crime prevention within the battery of law and order measures promoted by government. Initiatives in the government sector such as the Safer Cities Programme or efforts by national agencies such as Crime Concern are aimed at reducing crime through measures which involve partnership between local government (and/or the police) and local communities, usually in the form of voluntary bodies. Neighbourhood watch is the archetypal form of crime prevention project in this arena. With police and perhaps government support, it aims to motivate local communities into a higher level of protection against burglary. To counteract claims of ineffective use of resources, central government support for local crime initiatives is now clearly tied to monitoring. To be successful, projects need to stipulate how the impact of their activities will be assessed and to promise reviews at regular intervals. Local crime profiles are an essential part of such monitoring processes (Ekblom 1988). A similar logic applies to current initiatives aimed at diverting offenders from prison through community-based programmes (Home Office 1990).

The second additional support for local crime profiles emerges from the clauses of the Police Act 1984, which place a duty on the police to be more sensitive to the needs of the communities which they serve. One way this is achieved is by servicing local agencies such as Crime Prevention Panels or Community Liaison Committees with information pertaining to crime in their areas. Increasingly sophisticated information systems are being used for this purpose as well as by the police more directly to pinpoint particular local problems (Bailey 1989).

MODELLING LOCAL CRIME RATES

The strategy advocated here as an aid to the interpretation of local crime rates is a top-down model. The aim is to highlight levels of local crime by observing how they compare to national and regional patterns. The basic structure is a difference model, comparing actual local crime levels with those that would be expected if national or regional crime rates were applied to the local population. Differences are thereby generated between the actual number of recorded offences and the number that would be expected if national and regional rates were applied locally. Further illumination is provided by comparing rates for individual offence types to overall crime rates. In this way an attempt may be made to isolate whether a difference is due to a generally high level of crime in the area or to a particular problem with a specific type of crime. The simplicity of difference models is that all the effects add up, both for individual crime categories and as a whole. They also enumerate crimes and so work with numbers that have a ready interpretation.

The difference model

Table 3.1 outlines the form of the difference model which is specified in full in Figure 3.1. Essentially it is a variant of shift-share analysis (Randall 1973; Danson *et al.* 1980), which allows the difference between the number of crimes recorded in the local area and the number generated by applying national crime rates to the local population to be allocated to two distinct kinds of effects. One is the effect of differences in the overall crime rate and the other is the effect of the different mix of crime applying at the different levels. It thus allows a distinction to be made between a general local crime problem and a specific local crime problem, for example burglary or auto-crime. It further allows these distinctions to be related to a geographical source, local area within region, or region within national totals. In this way a crime problem may be seen to be a particular local problem or a reflection of a wider regional or national problem. Of course, based as it is on differences, the model deals only with relativities and is not geared to explaining factors exogenous to reported crime levels.

Table 3.1 Difference model of local crime rates

TOT DIFFS = LR + LM + RR + RM

where

TOT DIFFS = total local difference, i.e. the difference between the number of local crimes and the number generated by applying the national rate to the local population

LR = local crime rate effect, i.e. crime arising from differences between overall local and regional crime rates

LM = local mix effect, i.e. crime arising from difference in the mix of crime locally and regionally

RR = regional crime rate effect, i.e. crime arising from differences between overall regional and national crime rates

RM = regional crime mix effect, i.e. crime arising from difference in the mix of crime regionally and nationally

---

General form
TOT DIFFS = LR + LM + RR + RM
Formulae for crime type i

	Expansion
$TOT_i =$	$LT_i - LE_i$
$LR_i = RX_i - RE_i$	$[(LT_i - RE_i) - (LT_i - RX_i)]$
$LM_i = LT_i - RX_i$	none
$RR_i = NX_i - LE_i - RX_i + RE_i$	$[(LT_i - LE_i) - (RX_i - NX_i)$
	$\quad - (RX_i - RE_i) - (LT_i - RX_i)]$
$RM_i = RX_i - NX_i$	$[(LT_i - NX_i) - (LT_i - RX_i)]$

$LE_i = LP * NT_i/NP$
$RE_i = LP * RT_i/RP$
$RX_i = LT * RT_i/RT$
$NX_i = LT * NT_i/NT$

LT_i, RT_i, NT_i = local, regional, national number of crimes of type i

LT, RT, NT = local, regional, national crime totals

LP, RP, NP = local, regional, national populations

Figure 3.1 Specification of crime model

63

As an alternative to more traditional methods of summarizing geographical differences such as location quotients, similarity indices or Gini coefficients, the difference model offers distinct advantages in highlighting the elements of a particular local profile in terms of readily interpreted raw numbers. Lost is the direct ability to compare across a set of areas, but both approaches share a common root in their use of population-based ratios.

An application of the difference model

To illustrate the use of the difference model, it has been applied to the crime figures for a large housing estate on the outskirts of a medium-sized city in northern England. The estate consists mostly of council housing; the housing stock is generally new, little older than twenty years. The estate is regarded locally as having something of a crime problem, not particularly serious and certainly not as serious as some of the other problems faced by its inhabitants – lack of local job opportunities, low incomes, high bus fares, etc. On the plus side, there is a township shopping centre with local offices of many agencies. Schools are modern and relatively well-endowed. There is plenty of space and a degree of pride in belonging to the estate.

Table 3.2 shows that the overall level of crime on the estate is about double the national average. It might be thought that this constitutes more than the 'bit of a local crime problem' felt locally. The difference model, however, associates a substantial minor part of the excess of crime to regional rather than local effects. The estate is not so exceptional within a county that has distinctly above average crime levels. The relativities felt locally are therefore not so out of line with other similar places within the region.

Within the individual crime categories, four distinct groups of offences begin to emerge. Violence against the person and sexual offences both have local rates over three times the national norm. The bulk of the differences are local in origin, though a substantial minority of the effects are associated with regional mix. Thus the locally high rate of sexual and violent offences partly parallels the overall local crime problem and is partly tied to a specifically high rate for these offences in the county as a whole. The small local mix effects suggest that these offences do not constitute a specific local problem.

Burglary in a dwelling and the two auto-crime categories stand in

Table 3.2 A local profile of crime on a northern estate

Crime	Number of recorded offences	No. estimated by applying national rates	Total differences	Source of differences			
				Local effects		Regional effects	
				rate	mix	rate	mix
Violence against person	340	106	+234	+93	+37	+9	+95
Sexual offences	59	18	+41	+14	+12	+3	+12
Burglary dwelling	422	296	+126	+131	−5	+153	−154
Other burglary	563	253	+310	+241	−223	+2	+290
Robbery	16	21	−5	+3	+5	+17	−30
Theft from the person	14	21	−7	+4	0	+16	−27
Theft from shops	226	145	+81	+95	−82	+45	+24
Theft from a motor vehicle (MV)	686	417	+269	+192	+58	+207	−189
Theft or unauthorized taking of MV	394	246	+148	+98	+76	+139	−165
Other theft/handling stolen goods	1,211	469	+742	+335	+11	+114	+175
Criminal damage (value over £20)	626	288	+338	+176	+53	+100	+10
Fraud, forgery and other offences	117	105	+12	+51	−4	+50	−41
Total	4,674	2,386	+2,288	+1,433	0	+855	0

Notes: The crime types used here are those recommended for local profiles under the Safer Cities Programme. The regional effects are taken from the police force area in which the estate is located contrast to violence and sexual offences. Here the local rate is lower by comparison, though still above the national average (about 50 per cent higher). The differences in these categories are almost entirely local in origin, for at the regional level, the effects of county-wide high crime levels are counterbalanced by low specific county rates for these crimes. For the estate, therefore, these crimes do constitute a local problem.

For non-residential burglary and theft from shops, local crime rates are about twice the national average or about par for the estate. The difference model shows that these effects, however, are primarily regional in origin. In the case of non-residential burglary, the strongest effects come from regional mix, indicating a county-wide rather than local problem for this specific type of offence. In both these theft categories, the effect of locally high overall crime levels is counterbalanced by low specific rates for these offences.

The final group consists of other theft/handling stolen goods and criminal damage. Rates for the estate are slightly above average, about two to three times the national norm. Local effects again dominate but regional effects are also strong, particularly with other theft/handling having a large regional mix effect. These regional effects are similar to the previous group and perhaps point to a county-wide problem with theft in non-residential locations.

The 'crime problem' on this estate is therefore clearly far from uniform. Numerically local crime levels are driven by the large volume categories of other theft/handling stolen goods, criminal damage and theft from a motor vehicle and in all these cases the differences are predominantly associated with local rates. Numerically smaller but proportionately more exaggerated is the category of violence against the person. Burglary in a dwelling is relatively less serious a local problem.

The crime profile in context

Modelling local crime rates by the method outlined here is in itself an arid exercise. The interpretation of the figures needs to be set in the context of the social geography and physical layout of the neighbourhood. It is only against this information that a proper evaluation can be made (Locke 1990a: ch. 6). The crime profile needs a parallel community profile consisting not only of the statistical facts about the social structure, housing, land-use and facilities, but also about community feelings, the strength and weakness of community structures and the attitudes of residents towards crime and related issues. In our case study estate, the high rates for theft and auto-crime need to be set against a demographic structure currently emphasizing the crime-critical youth cohorts, a youth unemployment problem (though in fairness not in any way untypical of a northern English city) and a peripheral situation within the city. The combination of youth, poverty and geographical location is a firm foundation for explaining locally high crime rates. The fact that the rates for this estate are not exceptionally high indicates offsetting influences among which rank strong and active community groups, both voluntary and state-supported, and a neighbourhood centre which houses a wide range of services and local offices and which provides a focus for community life. Proper interpretation of local crime profiles thus involves not only setting a context for the crime rate itself through the modelling process but also setting a local context through an input of knowledge about the community.

THE ROLE OF LOCAL CRIME PROFILES

The preparation, interpretation and evaluation of crime profiles in their local context generates information useful in a broad spectrum of anti-crime activities (Heywood and Hall in Chapter 4). It may inform and enhance the process of policy review and the planning of schemes for intervention. It may make a direct input at an operational level in policing, crime prevention and offender-based services (Locke 1989a). The contribution of local crime profiles is achieved at different levels: in the identification or diagnosis of key targets both operationally and strategically, as an input to the development of a local response profile and in an ongoing process of monitoring and crime management.

Diagnosis

Once information has been gathered, some sense needs to be made of it. This involves diagnosis of identified problems (for example an increase in the rate of burglary on an estate) or of how a group of variables reveals a pattern that would require a range of responses by the police, probation services, courts and other agencies. The diagnosis of crime patterns and trends is a fairly well-established practice in police forces and other agencies such as probation services (Morgan 1985). However, it is often necessary to relate crime patterns to wider social and economic indicators in order to engage in policy development or operational reviews. In particular, demographic factors are of special importance in medium to long-term planning. Large youth cohorts in particular places, such as the estate used for illustrative purposes here, present a challenge across a wide spectrum of agencies concerned with crime and its consequences.

Smith and Laycock (1985) have used crime analysis to study burglary in a local area. Having identified a high rate of burglary their analysis proceeds to the kinds of houses that have been burgled, what kinds of goods stolen, the relationship between the location of the burgled houses and the area of residence of convicted burglars, the time of day the burglaries were committed and other factors. An analysis of this kind can provide pointers to crime prevention and policing strategies. In particular it can be used to target such measures more effectively. Broad patterns of crime can be used by local authorities in the planning of road transport services and in the layout of housing estates and parks (NACRO 1989). Street lighting schemes illustrate the benefits of an input of crime information (Painter 1990).

An overview of the crime prevention process is offered by Ekblom (1988). He clearly links the analysis and interpretation of data to the design of preventive strategies and indicates the need to move on to a continuous monitoring of crime. Various techniques can be used in the interpretation of crime patterns and in the diagnosis of specific problems. Mapping is an obvious technique to show the geographical dimensions of a problem and the spatial distribution of crime against the social, economic and topographical features of the district. Ekblom refers also to differentiation, recombination, creating new variables, frequency distributions, cross-tabulation and significance testing. Much of this work requires policy and planning reference points; for example, in order to evaluate conclusions for field agencies, some appreciation of their existing policy and of operational plans and activities is needed.

All crime management measures involve costs of one kind or another. Policy-makers and planners need to work within budgets and financial constraints. Individual householders and citizens need to evaluate the cost implications of their own personal crime security measures (Field and Hope 1989). Profiles should therefore present some information about the costs of crime to victims, to criminal justice agencies, the police and local authorities, and to other bodies that might be involved. There is a general need to increase the amount and use of economic and financial information in the evaluation, management and monitoring of crime problems (Locke 1989b).

From diagnosis to planning

Profiling is one method that can be used to draw conclusions about the impact of policing and prevention measures, to increase knowledge of crime and how it can be managed. Information is one of the key inputs to the process of planning local responses to crime (for a discussion of the general relationship between policy, planning, strategy and information in criminal justice systems, see Locke 1990a). Information can be used to evaluate the various alternatives and options that present themselves to police or probation officers, to magistrates and others. Most agencies involved with crime or justice accept the need for research and monitoring in order to improve their ability to respond effectively to crime. This takes place most usually within organizations but there is a growing realization that many of the responses that need to be made to crime need to take place between and among organizations – the local police division, probation office, district council, magistrates' court and so on. These inter-agency strategies pose a variety of problems arising from differences in organization, approach, statutory responsibilities and modes of operation. Profiles constructed for a particular locality provide a focus for inter-agency effort and could be used as a key to policy formulation. Profiles can therefore contribute to area policy and planning both to individual agencies and to any inter-agency bodies that exist, such as crime prevention panels, juvenile justice co-ordinating groups and so on. Considerable importance has been attached to effective inter-agency co-operation by the Home Office (1990) as well as by the police and local authorities in many areas of the country.

The response profile

Responses to crime are made by a variety of bodies apart from the individuals who are victimized. Criminal justice agencies are the 'front-line' agencies in crime responses but community groups, local authority departments, private security firms also contribute in a variety of ways. In any locality, crime will touch and provoke a response from a wide range of individuals, groups and organizations.

Some of these responses will be direct (such as arresting offenders) and some indirect (such as anti-poverty measures). It has been suggested that policy and planning processes require both a crime profile and a response profile (Locke 1990a). The response profile presents a picture of local reactions to crime and of the organized pattern of service that will affect crime and offending in the area. This suggestion broadens out the concept of the profile from being a technical statistical tool to being an instrument of policy planning.

Responses to crime generally fall into three main categories: prevention, diversion and 'correction'/'control'. Some responses are personal, some social and some environmental. Often there is little to be gained from dealing with crime as a whole. More often it is better to focus on particular types such as violence or property offences. In this way a matrix can be constructed to show how the responses to crime vary according to the nature of the offences. It also serves to limit the scope of a response to the resources and capabilities available.

Monitoring and crime management

The design and construction of profiles often takes the form of a research project (Unell 1983). The profile then becomes a snapshot of an area at one moment in time or over a short period. There are obvious dangers in this approach and it is much better to envisage an ongoing profile where data and information are gathered continuously as part of a strategic management process. Monitoring (as suggested earlier) is a well-established technique and crucial to policy formulation at all levels. It should be automatic therefore to update a profile once it has been piloted. The dynamic profile can reveal the interaction between crime and responses and thus feed an additional element into the policy and planning procedures of local organizations.

This leads to the concept of the strategic information system. It is currently fashionable to think of 'strategies' for crime management; strategic planning of response systems is becoming established as a means of controlling crime and this requires the design and implementation of strategic information systems. Dynamic profiling and modelling offer powerful tools to the strategic planner where one-off exercises soon lose their value as crime rates and operations change in a local area (Britton et al. 1988). Strategic policing requires a considerable body of information either to be supplied frequently or to be continuously available. Profiling and modelling enable information to be structured according to the use to which it is put and enhance its function of accurately, reliably and comprehensively informing decision-makers of what is happening and might happen in their area.

CONCLUSION

Local crime profiles are crucial to community-based anti-crime initiatives. We have attempted in this chapter to provide a model for evaluating the extent of crime in a community and have emphasized the importance of providing a context for the numbers that emerge. We have also indicated the contribution that the local profile can make to wider processes. We would end with a plea not to see the local crime profile as a one-off phenomenon, but as a part of an ongoing process of reviewing local crime patterns by and for the communities affected. We need longitude as well as latitude to pinpoint solutions to the crime problem.

REFERENCES

Bailey, S. (1989) 'Local crime', in *Law and Order Statistics Annual Conference Papers*, London: Statistics Users Council.

Bottoms, A.E. and Wiles, P. (1986) 'Housing tenure and residential community crime careers in Britain', in A.J. Reiss and M. Tonry (eds) *Communities and Crime*, Chicago: University of Chicago Press.

Britton, B., Hope, B., Locke, T. and Wainman, E. (eds) (1988) *Policy and Information in Juvenile Justice Systems*, London: Save the Children Fund.

Danson, M.W., Lever, W.F. and Malcolm J.F. (1980) 'The inner city employment problem in Great Britain, 1952–76: a Shift-share approach', *Urban Studies* 14: 193–210.

Davidson, R.N. (1981) *Crime and Environment*, London: Croom Helm.

Ekblom, P. (1988) *Getting the Best out of Crime Analysis*, Crime Prevention Unit Paper 10, London: Home Office.

Field, S. and Hope, T. (1989) 'Economics and the market place in crime prevention', *Research Bulletin* 28: 40–4.

Home Office (1989) *Criminal Statistics England and Wales 1988*, London: HMSO.

Home Office (1990) *Supervision and Punishment in the Community: A Framework for Action*, Green Paper, Cm 966, London: HMSO.

Locke, T. (1989a) *Area Profiling: Theory and Practice*, London: NACRO Young Offenders Team.

Locke, T. (1989b) 'The economic impact of crime in a local area: Leicester case study', unpublished paper, Urban Policy Studies, Leicester Polytechnic Business School.

Locke, T. (1990a) *New Approaches to Crime in the 1990's: Planning Responses to Crime*, Harlow: Longman.

Locke, T. (1990b) 'The ecology of urban crime', unpublished paper, Urban Policy Studies, Leicester Polytechnic Business School.

Mawby, R.I. (1979) *Policing the City*, Farnborough: Saxon House.

Morgan, P.M. (1985) *Modelling the Criminal Justice System*, Research and Planning Unit Paper 35, London: Home Office.

NACRO (1988) *Growing Up on Housing Estates*, Report by a NACRO Crime Prevention Committee Working Group, London: NACRO.

NACRO (1989) *Crime Prevention and Community Safety: A Practical Guide for Local Authorities*, National Safe Neighbourhoods Unit, London: NACRO.

Painter, K. (1990) 'Women's experience and fear of crime and the scope for public lighting as a means of crime prevention', paper given to the Leeds City Conference on Crime and Lighting, 6 June.

Randall, J.N. (1973) 'Shift-share analysis as a guide to the employment performance of West Central Scotland', *Scottish Journal of Political Economy* 20 (1): 1–26.

Reiss, A.J. (1986) 'Why are communities important in understanding crime?' in A.J. Reiss and M. Tonry (eds) *Communities and Crime*, Chicago: University of Chicago Press.

Smith, L. and Laycock, G. (1985) *Reducing Crime: Developing the Role of Crime Prevention Panels*, Crime Prevention Unit Paper 2, London: Home Office.

Unell, J. (1983) *Voluntary Action and Young People in Trouble: Volume 3, Local Surveys*, London: National Council for Voluntary Organizations.

4

IS THERE A ROLE FOR SPATIAL INFORMATION SYSTEMS IN FORMULATING MULTI-AGENCY CRIME PREVENTION STRATEGIES?

Ian Heywood, Neil Hall and Peter Redhead

One of the major problems in crime prevention is the formulation of the correct strategy for a given geographical location. This involves identifying the crimes which are most prevalent in the community; the social groups and geographical areas most at risk; the times when crimes are most likely to occur; and environmental and social factors which encourage crime. This chapter considers two approaches implemented within the Northumbria Police area to aid the targeting of crime prevention strategies. The first, 'The multi-agency approach', is organizational and tackles the need for a co-ordinated data-gathering and think-tank approach to crime reduction drawing upon the expertise of a wide spectrum of agencies that have an interest in local community affairs. The second approach is technological and involves a spatial database and geographic information systems (GIS) approach to the storage, management and manipulation of crime-related community data. For the purpose of this chapter it is necessary to make a clear distinction between spatial databases and geographic information systems. A spatial database is a relational database, in which a spatial reference has been used to relate items, whereas a GIS links database technology with the techniques of spatial analysis and computer cartography to provide a more versatile environment for the storage, retrieval, analysis and display of spatial information (an overview of database and GIS technology is provided in Everest 1986 and Burrough 1986 respectively).

The multi-agency and technical approaches were considered

compatible because they provide a framework within which information can be drawn together from a wide range of sources, related, and used to focus crime prevention policy. The approaches are also complementary because the multi-agency strategy tackles the problems of welding together organizations and decision-making structures, while the GIS and database technology provide the techniques for integrating different layers of community data. This chapter reviews the progress of these two approaches since their implementation on North Tyneside in 1986. The emergence of the multi-agency approach to crime prevention is described and the aims and objectives examined, focusing in particular on the Home Office Crime Prevention Unit's 'Five Towns Initiative'. The North Tyneside Crime Prevention Initiative (CPI) is then introduced and the reasons for the adoption of information technology and automated techniques of spatial data handling to develop a geographical crime database (GCD) are reviewed. Particular attention is paid to the spatial aspects of targeting crime prevention.

The use and limitations of the GCD and the potential of geographic information systems in crime prevention information management are considered. The problems, uses and benefits of the pilot Gateshead multi-agency geographic information system are evaluated and the future role of spatial information systems in crime prevention policy discussed.

THE FIVE TOWNS INITIATIVE

Historically, crime prevention has been mainly a police initiative, with individual forces adopting such strategies as target hardening (e.g. window lock campaigns); property marking; publicity (e.g. the MACPI Campaign); and the reallocation of personnel and resources to allow for community policing. Co-ordination and implementation of these strategies was left to the guidance of individual force crime prevention officers, who initiated programmes based on past experience and intuition. These projects tended to be all encompassing like the 'Thief is About' campaign, which was aimed at several different crime themes, from household burglary to car theft and initiated at a variety of different scales, from household mailing campaigns to displays in areas where a recent theft had taken place. In recent years, however, the importance and co-ordination attached to crime prevention programmes have taken on new dimensions as traditional policing methods have been placed under considerable pressure by the increasing levels of crime within the community. The attitude that 'prevention' is as important as 'cure' has now become a common theme in many police forces and is given emphasis by the Home Office (e.g. Richardson and Eden 1987).

In 1984 the potential value offered by adopting a crime prevention approach was realized by the government with the implementation of the Home Office 'Five Towns Initiative' (Home Office 1984). This initiative put in place five pilot crime prevention schemes located at Bolton, North Tyneside, Croydon, Swansea and Wellingborough. These areas were chosen to provide both a wide geographical spread and to ensure that the initiatives covered a comprehensive range of social conditions and problems. The initiatives were allocated funding for an eighteen-month period, during which time their remit was to target accurately local crime problems, to attempt to reduce their occurrence, and to alleviate their impact within the local community. The projects were to establish multi-agency task forces, involving not only those agencies traditionally associated with crime prevention (primarily the police) but also those agencies with a vested interest in the community. This 'multi-agency approach' aimed to be a community-driven response to crime reduction. For example, local authorities were encouraged to implement design changes to their properties to create a less vulnerable environment; residents were encouraged to participate in neighbourhood watch schemes; and the police were urged to adopt a more community-based approach to crime prevention, with an increase in the number of local permanent beat officers and the distribution of local crime newsletters. In addition, other public and private sector bodies were encouraged to promote policies related to crime prevention. These included inputs from social services, probation, education and even past offenders.

The Home Office Crime Prevention Unit suggested that the CPIs should have three main objectives. The first focused on the need to reduce crime within the community, and it is in this area where the traditional techniques such as 'target hardening' (see Laycock 1988), complemented by those from the newer school of 'designing out crime' (Clarke and Mayhew 1980) have been employed. The second aimed at reducing the fear of crime prevalent within certain community groups, particularly elderly people. Formulating policy in this area generally centred on the need to increase community awareness of crime-related issues. However, it was appreciated that careful attention had to be paid to prevent alarming the community and actually enhancing community fears, as has been found to be the case in the excessive use of security devices (Winchester and Jackson 1982). The final objective the Home Office saw for the CPIs was more broadly based and associated with enhancing the local community spirit and environmental awareness, based on the belief that a community which considers itself to be a community creates an environment in which the criminal element is less likely to be active (Shapland 1988). Community spirit was encouraged via the now familiar vehicle of neighbourhood watch schemes and other innovative projects such as local crime prevention newsletters.

A management model was provided as a guideline by the Home Office Crime Prevention Unit and consisted of two main elements. First, a full time co-ordinator seconded from one of the local agencies: in the case of North Tyneside this was a chief inspector from Northumbria Police. The second part of the model was comprised of a steering committee, made up of representatives from the collaborating agencies, to oversee the development and progress of the initiative. The major role of the steering group was to guide the initiative in the development of CPI projects and help to establish a mechanism to acquire information about the community. In addition, it was appreciated that a third tier would be required to aid the co-ordinator in developing a community crime prevention strategy. During the initial phase of the 'Five Towns Initiative' the Home Office Crime Prevention Unit gave only limited advice on how this tier might be established as it was appreciated that each initiative would have access to a different resource base. The nature of this resource base would, in turn, determine the character of the initiative's operational framework. On North Tyneside, provision was made for a project team comprising of four elements; research, administration, secretarial and fieldwork. This project team was financed by funds made available to the initiative from sources such as the Community Programme, Northumbria Police and the Urban Programme.

THE NORTH TYNESIDE CPI AND THE DEVELOPMENT OF A CRIME PROFILE

The North Tyneside Crime Prevention Initiative (CPI) set up under the Five Towns Initiative programme was aimed at two high-crime areas on the north-east edge of Newcastle upon Tyne, encompassing the post-war council estates at Longbenton and the new town at Killingworth. The project adopted the management 'blueprint' suggested by the Home Office Crime Prevention Unit. The co-ordinating role was undertaken by a chief inspector seconded to the Home Office by Northumbria Police and a steering panel comprised of senior representatives from the relevant local agencies. The initial objective of the initiative was to develop a crime profile of the project area. This profile needed not only to outline the characteristics of the crimes most prevalent within the community but also to identify the most vulnerable geographical areas. The North Tyneside initiative saw the role of the crime profile as a data resource which would help identify the structure of crime within the project area and, in turn, act as a catalyst to stimulate the steering panel to focus their attention on possible solutions. In the first instance, it was anticipated that the primary data layer of this profile should be police information on reported crime. These official statistics were used to identify the main target crimes of burglary offences (directed at both residential and commercial premises) and auto-crime (including theft of, and theft from, motor vehicles).

Unfortunately, reported crime statistics gave an incomplete picture. A graphic example of this was the official statistical pattern of vandalism which, in Killingworth, indicated that 84 per cent of offences were committed against private property. It did not take any in-depth thinking to recognize the falsity of this picture. The extent of graffiti, broken bus shelter windows and smashed street furniture told a different story. Likewise, crime surveys, most notably the British Crime Survey (Hough and Mayhew 1983), had indicated that there were similar failings in most other reported crime classifications (Mayhew *et al.* 1989). This missing or 'dark figure' associated with crime must be assessed if community crime prevention programmes are to be targeted accurately. As a result of these gaps in police statistics, the North Tyneside CPI decided to complement the preliminary crime profile by drawing on surrogate information collected for a variety of other purposes by participating agencies. For example, recorded incidents of damage to educational property by educational authorities and information on damage to street furniture from the local authority departments associated with public works and engineering services. In a large number of cases, however, particularly with crimes to the individual, no surrogate data sources existed. Therefore, a series of detailed residential, commercial and offender surveys were designed and implemented to provide information on the dark figure associated with unreported crime in the community. These information-gathering exercises were also seen as a way of involving the community in the decision-making stages leading up to the formulation of local crime prevention policy. They also formed a mechanism through which to encourage the growth of community spirit.

THE DEVELOPMENT OF THE GEOGRAPHICAL CRIME DATABASE

The North Tyneside Initiative decided to use the information provided by the crime profile to develop a micro-based geographical crime database (GCD). The role of the GCD was to provide a comprehensive source of information which could be interrogated at various degrees of spatial resolution to identify the characteristics of crime within the community; current crime prevention methods (for example the use of video surveillance by commercial firms); the variety of environmental factors important in terms of crime prevention (for example the presence or absence of street lighting); demographic characteristics; and community perception. A decision was taken to give this database a geographical component because it is recognized that crime does not occur randomly across an area and that very often the nature of the geographical location may actually enhance the likelihood of the crime being committed. Several authors, for example, Coleman (1985), have commented on how the character of the built environment can influence crime. The dark unlit aerial walkways and alleyways of the Killingworth Towers estate on North Tyneside were a prime example of this, providing fast and easy access for the 'would-be' burglar as well as a quick escape route outside the reach of the vehicle-bound emergency services. Wilson and Kelling (1982) provide further examples of how 'incivilities', such as vandalism to street furniture, graffiti and boarded-over windows, can influence the geography of crime throughout a region. Moreover, the perceptions of the community are influenced by the nature of the locality. Wilson and Kelling (1982) suggest that the crime fears of a community's residents are likely to be higher in those areas showing greater 'incivilities'. Therefore, relating people to places can help both in problem targeting and identifying solutions.

The North Tyneside CPI used three scales of geographical referencing to permit identification of crime prevention problems at different spatial levels; estate name, street name and postcode. The use of postcodes also permitted spatial units to be created at a variety of intermediary levels encompassing different groups of streets or residential blocks. Accurate grid referencing was not used in the development of the initial GCD because of the problematic and time-consuming nature of obtaining such data. Primarily, only two components of the crime profile, the residential and commercial surveys, were used in the development of the GCD. This was due to the fact that the majority of information resources that were made available to the initiative by the collaborating agencies, though relating information to geographical locations did not do so with any degree of spatial accuracy. The most obvious disappointments in this area were police crime incident reports which used a variety of *ad hoc* methods for the locational referencing of an incident, from the occasional use of the postcode to approximate descriptions of area where the crime or attempted crime was committed. Encouraging the participating agencies to adopt a more rigorous approach to tagging data with an accurate spatial reference is one of the issues that is currently being pursued by the initiatives. Northumbria Police have now recognized the benefits that postcoding of crime incident data could provide for using these data in crime forecasting, crime pattern analysis and strategic planning.

A further factor which limited the development of the initial GCD was the lack of access to information from the other participating agencies in a readily accessible computer readable form, either because agencies stored information in a paper-based form as tables, text or map information, or on larger mainframe systems from which it was difficult to gain access to the data because of problems of compatibility or confidentiality.

The pilot GCD used on North Tyneside was developed using relational database technology and interfaced with graphics and statistical analysis software to allow for the analysis and visual display of data (a detailed description of the system is provided in Heywood and Hall 1987). At the time of developing the GCD insufficient funds were available to make provision for a mapping component to the system. This is considered to be one of the primary drawbacks of the initial system. The structure of the GCD consisted of a series of different hierarchical data layers which were related by both a unique spatial and record reference (see Figure 4.1). A turnkey environment was programmed for data input to enable field-workers to enter survey data via a user-friendly interface designed to minimize the entry of bad data. Steps were taken to programme a similar environment for data retrieval, analysis and display but this proved difficult since it was not possible from the outset to anticipate the type of interrogative questions the steering group might wish to ask. It was decided to keep the querying component of the database flexible until a broader spectrum of information requirements was identified.

USING THE GCD

The use of the GCD in formulating crime prevention policy has been discussed in some detail in Heywood and Hall (1987). Its primary use was as a tool to aid in the preparation of the crime profile and as a decision support system to help in a variety of crime prevention areas, from the identification of geographical distribution of target crimes through to evaluating the likelihood of support for neighbourhood watch schemes. The example cited in Heywood and Hall illustrating how the GCD was used as a decision-support system is summarized here by way of example. Based on police information, it was recognized that ground-floor flats posed an attractive target for burglars due to ease of access through side entrances and windows. By browsing the residential survey data contained within the GCD it became apparent that, while there were a number of burglaries to ground-floor flats in the project area, there were substantially more entries made into first- and second-floor flats than anticipated. Moreover, when the database was queried in more detail, these particular crime incidents were clustered geographically on the Longbenton council estate. A further examination of the GCD revealed that the backdoor was the entry point in virtually all cases. Once the GCD had been used to identify this problem, field-work was undertaken on the ground to examine more closely the built nature of these premises. This revealed a very open balcony structure to the rear of all the second-floor flats, which was an ideal environment for permitting unobstructed access for a potential burglar. Once this problem had been brought to the attention of the steering panel a policy was devised which involved enclosing these exposed balconies. This had the dual effect of removing the burglary access point and enhancing the living space of the dwelling.

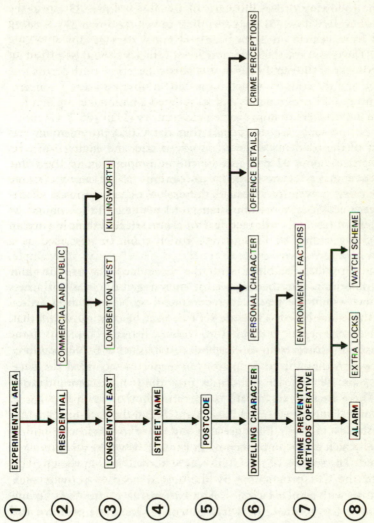

Figure 4.1 The structure of the North Tyneside geographical crime database

① EXPERIMENTAL AREA

② RESIDENTIAL — COMMERCIAL AND PUBLIC

③ LONGBENTON EAST — LONGBENTON WEST — KILLINGWORTH

④ STREET NAME

⑤ POSTCODE

⑥ DWELLING CHARACTER — PERSONAL CHARACTER — OFFENCE DETAILS — CRIME PERCEPTIONS

⑦ CRIME PREVENTION METHODS OPERATIVE — ENVIRONMENTAL FACTORS

⑧ ALARM — EXTRA LOCKS — WATCH SCHEME

Throughout the eighteen months of the pilot project the GCD was used in similar ways to the example described above to help focus crime prevention strategies throughout the North Tyneside target area. At the end of the trial phase of the project official police statistics for the area showed a reduction in all of the target crimes to the following order: burglary of dwellings down 23 per cent; other burglary down 51 per cent; theft of vehicle down 23 per cent; theft from vehicle up 22 per cent; criminal damage down 18 per cent. The rise in the theft of items from vehicles should be placed in the context of the rapid rises in this particular crime both across the region and the country which indicated an obvious area of concern for future CPI programmes and stimulated a national campaign for car manufacturers to improve vehicle security (Hall 1987). Of more interest, perhaps, are the details that the GCD provided on the views of the residents when re-surveyed after the eighteen-month trial phase. Some 92 per cent of the community considered the initiative successful and wished it to continue, 35 per cent had taken some crime prevention measures themselves as a direct result of the project's activities; more importantly, 61 per cent of the original 80 per cent of residents who said fear of crime affected their life-styles had changed their outlook to one which could be described as a proper concern about crime.

These results, combined with the visual improvements to the communities within the target area, encouraged the local authority districts within the region to recommend Northumbria Police that the initial North Tyneside CPI project be continued and that further schemes be extended to other districts. Out of these discussions a partnership approach was established with Northumbria Police providing the co-ordinator (an inspector seconded for three years) and the district authorities providing the accommodation, etc. There are now six initiatives operating in the North East based in North Tyneside, Central Newcastle, Sunderland, South Shields, Gateshead and Blyth. Plans are also underway to devise a simplified model which can be implemented in rural areas where resources are limited. The success of the GCD was acknowledged in the upgrading of the CPI programme by the agreed need to provide each initiative with similar facilities. At a national level, the Five Towns Initiative has formed the basic blueprint recommended by the Home Office for those areas in the UK which wish to adopt a multi-agency approach. In addition, the Five Towns Initiative model forms the basis for the current Safer Cities project, which is aimed at inner city areas throughout the country.

THE GATESHEAD MACPI SPATIAL INFORMATION AND MAPPING PROJECT

The North Tyneside GCD was developed primarily as an experimental tool to examine whether modern automated methods of data handling could be used successfully to help target crime prevention strategies. The development of the GCD was never seen as being a comprehensive stand alone information system. However, its successful use by the North Tyneside CPI indicated that there would be obvious benefits to be gained from developing further the concepts underpinning the GCD into a more comprehensive information system. The first stages of this are now in evidence, as all of the new initiatives in the region have pursued a co-ordinated approach to their data-handling strategy so that information can be exchanged and compared between initiatives as required.

However, there were three major problems with the North Tyneside GCD. The first was the inability of the system to present spatial information in map form. Locational information about factors likely to influence crime prevention required substantial manipulation by the researcher and the replotting of information manually on maps before it could be conveyed to the steering panel. The second was the unfriendly nature of the database interface used to retrieve information. The third problem was one of integration. For example, it was impossible to integrate data from the GCD with other spatial information about the community, such as demographic profiles, unemployment statistics, or the proximity of incidents to other spatial features such as major public transport routes.

The first of these problems could have been solved by the use of computer mapping software to enhance the presentation of crime-related information. In fact, such a system had been developed by the Home Office Scientific Research and Development Branch in conjunction with Staffordshire Police Force (Home Office 1987). However, while this would have helped in the visual presentation of information, it would not have helped to overcome the problems associated with ease of use or those of data integration. Therefore, at this stage, a decision was taken to experiment with GIS software.

GIS systems are concerned with the storage, manipulation and management of information that relates data to places. These systems are not new in concept but offer a new way of handling and visualizing data. They combine the existing power and flexibility of established database management techniques with the concepts of automated mapping, and provide an added component which permits the analysis and manipulation of topographic features such as areas (police beats), points (individual incidents) and lines (roads) and their associated characteristics (attributes). These systems make use of the developments in modern information technology, in particular processing power and high-quality display facilities. GIS systems harness this power to permit the rapid analysis of large social and physical data sets utilizing the dimensions of time and space to provide new perspectives on existing data and generate new information. This information can then be displayed as maps, graphs, diagrams, tables and text to help target decision-making strategies and disseminate information.

It was in the analysis of crime prevention data and especially their effective explanation and display to a wide and varied audience that it was considered a GIS system would prove invaluable, providing advantages over the GCD. For example, capturing the particular target area in computer map form would in future permit data display at street or even house level thus making the problems of relating research results to actual geographical locations simplistic. In addition, access to data in this manner from microcomputer terminals would allow all interested parties from beat police-officers to chief local authority architects to have quick answers to problems shown in an instantly recognizable map form. The possible benefits of this approach can be shown by relating police burglary details for the target area (including burglary *modus operandi* data) to local authority home security work, thus helping to reveal which types of door locks, doors and windows appear to be deterring criminals most effectively. Such data relations can also show how effective neigbourhood watch schemes are over time and identify whether changes in police beat routes have been accompanied by a change in crime patterns. The GIS approach provides an easily understandable display of what can be complex spatial temporal analyses of use to many agencies and acts as a focal point, drawing these diverse bodies together by always relating results to the community at risk.

While it would appear that GIS technology provides the answer to developing an information system from a multi-agency point of view, a number of problems need to be mentioned.

1 Micro versions of this type of software have only recently been made available, and therefore choice and reliability are both limited and questionable.

2 Because the software is innovative programs are expensive, at least an order of magnitude above the cost of other more established applications software, such as spreadsheet and database systems.

3 The hardware needed to run them is expensive, top-range micros being required, and quality plotters and monitors essential if the real benefits are to be gained. In addition, at present only a limited amount of map information is in digital form and this is most likely not at the resolution required by initiatives for strategic planning purposes. Therefore, data capture hardware (such as a digitizer) and software maybe required to turn map-based information into digital map form.

4 In the past, one reason GIS systems have failed is because they have missed their primary objective, which is to enhance the ability of the policy-maker to make decisions, by being developed externally to the organizations which required to make use of them.

Since the majority of these problems also represent substantive costs, which none of the initiatives were in a position to meet, it was decided to implement a pilot project to evaluate whether GIS technology might be a cost-effective mechanism for developing a crime prevention information system.

The evaluation project was undertaken by the North Eastern Regional Research Laboratory based in the Centre for Urban and Regional Development Studies at Newcastle University in conjunction with Gateshead MACPI. The initial phase focused on the development of a demonstrator system covering a part of the project area and incorporating a subset of information from the initiative's GCD. The geographical area selected for the project was the Old Fold estate in Gateshead, which is comprised of an area of 1930s council housing with high crime levels and severe social problems, in particular high unemployment. The estate has approximately 400 dwellings and includes two junior and infant schools, a community centre and some light industrial property. The data subset chosen from the GCD included: burglary and vandalism offences from 1 April 1987 to 31 March 1988; the security measures of each house; and the crime concerns of the estate's residents.

The aim of the pilot project was to examine whether the additional flexibility of the GIS software would provide advantages over the established relational database software. As already illustrated, it was anticipated that the primary advantages would be in the visual presentation of spatial information to those responsible for formulating crime prevention strategies at an operational level. The following five objectives were identified as required of the project.

1 The acquisition of the relevant map information and crime prevention data.
2 Development of a data archive from which information could be retrieved and displayed when required.
3 Development of the necessary data analysis infrastructure to convert data contained within the GIS into information which could be used to formulate crime prevention policy.
4 Development of a user-friendly mechanism for displaying data in textual, tabular, graphical and map form.
5 Development of a user-friendly interface which would permit the untrained user to enter, analyse, retrieve and display information.

The demonstrator system was implemented on an IBM PS2 Model 80 microcomputer using PC ARC/INFO GIS software (Dangermond 1986). The chosen area was captured from the 1:12,500 Ordnance Survey map of the estate. The map detail was simplified to include only the road network and built environment. Each of the premises was given a unique label which allowed it to be linked to the subset data from the initiative's GCD. The postcode was not used, because it was considered to be too general a locational reference at this scale. No individual personal information linking a resident to a given address was recorded within the system, primarily for data protection reasons. A schematic diagram of the system is provided in Figure 4.2.

The system uses a computer map of the estate as the key to accessing the data from the GCD. Either the entire estate or particular areas down to individual premises can be selected by use of a zoom facility. When the user has chosen the geographical resolution at which to interrogate the system, there is a facility for browsing through the available information. Based on the concepts of the GCD this can either incorporate displaying all the information or displaying information for only those areas which meet the criteria established by the interrogator. For example, the user could ask to display all occurrences of burglary for the chosen area, or display only those incidences of burglary which had occurred at dwellings where target hardening measures had been employed. The

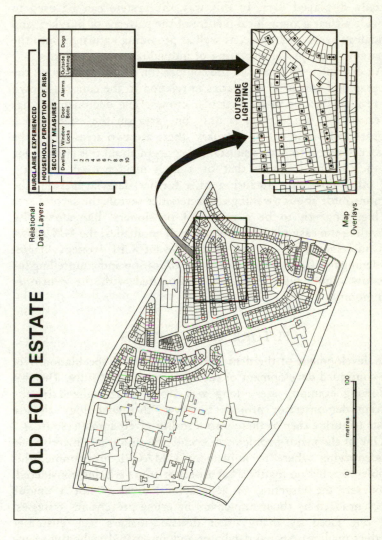

Figure 4.2 Schematic diagram illustrating the structure of the Gateshead geographic information system (GIS) demonstrator

system allows for the display of retrieved information in either the traditional tabular form associated with the GCD or in map form, using a variety of cartographic techniques, such as choropleth mapping to represent the different significances attached to the visually displayed data. In this way the system can be used to provide a clear geographical picture of the patterns of burglary and vandalism across the estate, as well as providing an insight into the apparent geographical clustering of individual crime concerns.

An example of its use is in the comparison of information on the geographical spread of burglaries in relation to the concern shown by residents about these particular crimes. The demonstrator has revealed that, while there is only one area on the estate which appears to be hard hit by burglary, there are two areas where the residents appear to have a substantial concern about this crime. One of these coincides with that of a high incidence of burglary, the other does not. Inclusion of a further data layer from the demonstrator about dwelling characteristics reveals the second area of high concern to be a group of pensioners' bungalows. By providing the easy identification of such anomalies, the GIS allows for implementation of the most relevant CPI strategy: target hardening for the high burglary areas; education and counselling for the low crime, high fear, sector associated with the pensioner community.

FUTURE APPLICATIONS

The development of the demonstrator has led to speculation over the continued development of such systems in the future. The two following examples are a long way from being realized by the current demonstrator project, however, the technology already exists to turn either of these into a reality given ample resources.

One of the major problems in crime prevention at a macro-scale is identifying where to implement a MACPI programme: this problem applies no matter what aspect of crime is being considered. However, the targeting of geographical areas is not a unique problem faced by those implementing crime prevention strategies, but one faced by many other decision-makers (e.g. customer product mailing). Spatial databases and automated techniques have already been developed to provide mechanisms for targeting strategies at the optimum geographical areas (Openshaw 1987). These strategies combine new spatial data sets with existing geographically referenced data sources, such as postcode and small area statistics (e.g. census data). An example of this type of product is 'Super Profiles', a product developed by the North East Regional Research Laboratory, which allows consumers' product mailing to be targeted based upon the social profile of an area (Openshaw 1987; 1988). A sideways step from this type of system could be the development of a spatial information system linking both police statistics and social data sets for a region, thus allowing the fine tuning of individual MACPI programmes to meet the requirements of respective localities. A system developed on this scale could also be used to monitor the overall effectiveness of programmes throughout an area and perhaps provide substantial evidence to illuminate the degree of crime displacement that these initiatives might cause. In this manner, the system could be used to guide initiatives as to when they might need to reconsider their target crimes.

A further step forward in the future, from using GIS techniques to target crime prevention strategies, might be the development of a geographical, on-line, crime prevention information system. This would tackle one of the major problem areas in crime prevention: the dissemination of correctly targeted advice. Information could be made available, either via existing telecommunication systems (for example, Oracle or Teletext) or by the provision of computer access terminals in public places such as libraries and community centres. The system could provide current details on criminal activities in specific geographical areas; requests for information in the *Crime Watch* mode; and local information on crime prevention strategies. In addition, the system could be used to capture data on a local community's views about crime prevention which, in turn, could be a useful guide in formulating new initiatives.

CONCLUSION

The rapid spread of MACPI throughout the Northumbria Police Force region is a clear indication that these initiatives are, at least in the short term, considered to be a cost-effective method for tackling community crime. Blagg *et al.* (1988) have commented on the potential problems of inter-agency wrangling, however, this appears to have been avoided in the North East by careful co-ordination. The success of the programme is no doubt due to accurate problem targeting, which in turn has been helped by founding crime prevention policy on spatially referenced databases.

Only time, in association with continued monitoring of the target areas, will reveal whether the initiatives are to have a long-term impact. A preliminary indication that this may well be the case has been shown on North Tyneside, where the scale of the pilot project has been reduced, but crime has continued to fall. Whether the complete withdrawal of an initiative from an area would lead to a reversal in this trend has not yet been tested. There are, of course, the obvious questions to be raised about crime displacement both in terms of geographical location and crime type. One clear example of this later form of displacement has been the rapid increase in thefts of items from vehicles, which has run in parallel with the growth in neighbourhood watch campaigns, where the primary focus of attention has been on dwelling security. Changes in crime trends is an obvious area to which MACPI must pay particular attention in the formulation of future crime prevention strategies.

Undoubtedly, the concept of thorough analysis using spatially referenced databases has been proved sound, as the examples from North Tyneside illustrate. However, the added value of adopting GIS technology to make provision for a crime prevention information system needs to undergo further evaluation. The Gateshead pilot project identified a number of advantages stemming from the adoption of a GIS approach, primarily associated with the manner in which a wide range of community crime related information can be assembled, processed, and communicated rapidly to the decision-maker. At the present time, those involved with the project consider the major factors which will limit the adoption of GIS are the high set-up costs and limited availability of data at community level in a digital form. The examples on possible future developments presented at the end of this chapter also raise fundamental moral questions about the potential dangers of linking personal data sets. Problems relating to confidentiality and the need for personal data aggregation are very often dismissed all too lightly (Woodhead 1987) and these need to be given thorough consideration before pursuing the development of any operational crime prevention information systems.

Acknowledgements

This project could not have been completed without the support of both the North Tyneside and Gateshead MACPI organizations who provided both valuable information and advice. In addition, two individuals deserve special recognition: Richard Pulsford for his technical work on the development of the GIS demonstrator and Gustav Dobrzynski for vastly improving upon the cartographic output of the system.

REFERENCES

Blagg, H., Pearson, G., Sampson, A., Smith, D. and Stubbs, P. (1988) 'Inter agency co-ordination: rhetoric or reality', in T. Hope and M. Shaw (eds) *Communities and Crime Reduction*, Home Office Research and Planning Unit, London: HMSO.

Burrough, P.A. (1986) *Principles of Geographical Information Systems for Land Resources Assessment*, Oxford: Clarendon Press.

Clarke, R.V.G. and Mayhew, P. (1980) *Designing Out Crime*, London: HMSO.

Coleman, A. (1985) *Utopia on Trial*, London: Hilary Shipman.

Dangermond, J. (1986) 'ARC/INFO – a modern geographic information system for large spatial databases', in B.K. Opitz (ed.) *Geographic Information Systems in Government*, vol. 2, Hampton, VA: A. Deepak.

Everest, G.C. (1986) *Database Management Objectives: System Functions, and Administration*, London: McGraw-Hill.

Hall, N. (1987) 'Cutting Crime on the Tyne', *Police Review*, February: 274–6.

Heywood, D.I. and Hall, N. (1987) 'Crime prevention: the value of spatially indexed databases', *Newsletter of the British Urban and Regional Information Systems Association*, September: 11–13.

Home Office (1984) *Crime Prevention: Circular 8/1984*, London: Home Office.

—— (1987) *CAP – Crime Analysis Package: USER GUIDE*, V2.0, Scientific Research and Development Office, London: Home Office.

Hough, M. and Mayhew, P. (1983) *The British Crime Survey: First Report*, Home Office Research Study 76, London: HMSO.

Laycock, G. (1988) 'Property marking as a deterrent to domestic burglary', in K. Heal (ed.) *Situational Crime Prevention from Theory into Practice*, Home Office Research and Planning Unit, London: HMSO.

Mayhew, P., Elliott, D. and Dowdes, L. (1989) *The 1988 British Crime Survey*, Home Office Research Study 111, London: HMSO.

Openshaw, S. (1987) 'Analysing and exploiting client data by creating a geographical marketing information and modelling system', *North East RRL Research Report*, 2, University of Newcastle upon Tyne.

—— (1988) 'A look in to the future of demographics', *New Generations*, summer issue, Liverpool: Credit and Data Marketing Services.

Richardson, B. and Eden, A. (eds) (1987) *Crime Prevention News* 3, London: Home Office.

Shapland, J. (1988) 'Policing with the public' in T. Hope and M. Shaw (eds) *Communities and Crime Reduction*, Home Office Research and Planning Unit, London: HMSO.

Wilson, J.Q. and Kelling, G.L. (1982) 'Broken windows: the police and neighborhood safety', *Atlantic Monthly* 255: 29–38.

Winchester, S. and Jackson, H. (1982) *Residential Burglary: The Limits of Prevention*, Home Office Research Study 74, London: HMSO.

Woodhead, K. (1987) 'The Chorley Report: a first reaction', *Newsletter of the British Urban and Regional Information Systems Association*, September.

Part II

CRIME AND POLICING: SPATIAL PATTERNS AND SOCIAL PROCESSES

5

WHERE CRIMINALS LIVE
A study of Budapest
János Ladányi

This chapter summarizes results of research into the spatial segregation of social groups in Budapest and explores in some detail the residential segregation of imprisoned offenders in the city. It also assesses police claims that there are areas of Budapest in which there are high concentrations of offenders, using data on the place of residence in Budapest of those in prison in 1979. One focus is a comparison of these concentrations of offenders with the most disadvantaged group in Hungarian society, the gypsies. The chapter studies issues not previously researched in Hungary.

RESIDENTIAL SEGREGATION IN BUDAPEST

Since the early 1900s, urban geographers and sociologists have been interested in the residential segregation of social groups in Budapest. Using some of these early studies in conjunction with more recent data it has been possible to trace the changing patterns and degree of residential segregation from the 1930s through to 1980 (Ladányi 1988; 1989). From a relatively high degree of spatial segregation between low- and high-status groups in the 1930s, there was a significant decline in the 1940s, when the Communists came to power in Hungary. This decline in residential segregation slowed down in the 1950s, since which time there has been a steady increase in the segregation of low- and high-status groups in Budapest. These changes in residential segregation parallel major social and political and economic changes in the country (Ferge 1969; Andorka 1982; Kolosi 1984). Thus, for example, the moderation of the extent of spatial segregation among different social groups in the 1950s can be linked to the disintegration of the pre-war capitalist regime and the underdeveloped condition of the state socialist

95

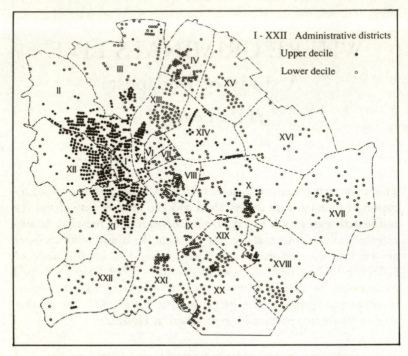

Figure 5.1 Semi-skilled and unskilled economically active workers in Budapest 1980

regime which was to replace it. Although the pre-war structures of inequality had disappeared, the new forms of inequality and privilege which were to emerge under the state socialist regime had yet to become fully developed. Indeed, it was only by the 1970s that these new forms of inequality had become manifest in the patterns of residential segregation in Budapest.

Turning to the characteristics of the spatial segregation of low-status social groups in Budapest, recent research has revealed that this segregation is significantly different to that identified elsewhere by western commentators. Based largely on the North American experience, much of the literature on urban segregation suggests that high levels of segregation exists for groups at both the top and the bottom of the social hierarchy. However, in Budapest, as in many other European cities, it appears that the degree of segregation of social groups at the bottom of the social hierarchy is significantly lower than that for groups at the top (see Ladányi 1989). The reasons for this lies in the power exercised by different social groups in cities. While high-status groups have sufficient economic and political power to achieve considerable territorial control over areas of a city, choosing where to live and excluding other social groups, low-status groups can live only in those parts of the city where nobody else wants to live. Indeed, as Figure 5.1 shows, the high-status groups are segregated in a few, large, exclusive housing areas, whereas low-status groups are segregated in many, small, mixed land-use areas. An important exception to this pattern is, however, the spatial segregation of the gypsies (Ladányi 1989). There are areas of Budapest which are almost exclusively inhabited by gypsies. As Figure 5.2 shows, almost 50 per cent of the gypsy population lives in the slum areas of inner Pest, which are dominated by state-owned flats, whereas no gypsy families live in

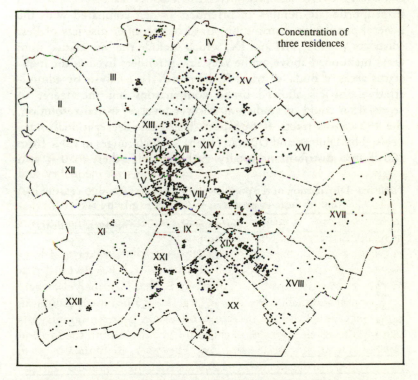

Figure 5.2 Residential distribution of gypsies in Budapest 1980

the high-status districts on the Buda mountains, where most housing is in the form of private flats.

THE RESIDENTIAL DISTRIBUTION OF IMPRISONED OFFENDERS IN BUDAPEST

The sample of offenders studied in this project (2,284 individuals) consisted of those serving sentences in Hungary on the 31 October 1979 who were over 18 and who gave Budapest as their permanent or temporary address (but not those living in workers' hostels or reformatory schools). Preliminary analysis of the age and educational status of these offenders revealed that they were generally much younger and less educated than the population of Budapest as a whole (Tables 5.1 and 5.2).

Turning to the spatial distribution of the imprisoned offenders, the initial analysis involved plotting the residence of the offenders in Budapest's twelve administrative districts. As Table 5.3 shows, a disproportionate number of offenders (when compared with the general population distribution) lived in the inner districts of Pest (districts VI, VII, VIII and IX) which include the low-status, slum areas mentioned above, while very few offenders lived in the high-status areas of Buda (districts II, XI and XII). Analysis by administrative districts allowed broad comparisons but the degree of segregation could be studied at a lower level of spatial resolution, the 490 census tracts of Budapest. Using the segregational index devised by Duncan and Duncan (1955) – which ranges in value from 100 if two distributions totally exclude each other, to 0 if two

Table 5.1 Distribution of the population in Budapest (over 20 years of age) and of those in prison in 1979 by age group (%)

	Imprisoned	Budapest residents*
20–24	19.1	8.8
25–29	28.8	11.3
30–34	18.1	10.1
35–39	12.9	9.5
40–44	8.0	7.6
45–49	5.5	8.3
50–54	4.0	9.0
55–59	2.0	9.1
>60	1.6	26.3
	100.0	100.0

Source: 1980 census Vol. 1. Data of Budapest, II, pp. 8–9.

Table 5.2 Distribution of the population in Budapest (over 20 years of age) and of those in prison in 1979 by their educational level (%)

	Imprisoned	Budapest residents*
Did not finish primary school	13.4	22.9
Finished primary school	46.8	28.8
Did not finish vocational training school	4.0	0.3
Finished vocational training school	18.1	7.9
Did not finish other secondary school	2.6	1.5
Finished other secondary school	11.1	24.1
Did not finish higher education	0.3	2.0
Finished higher education	2.2	12.5
Lack of data	1.5	0.0
Total	100.0	100.0

* Source: As for Table 5.1

Table 5.3 Residential distribution of imprisoned offenders in 1979 and of the population as a whole in Budapest (%)

District	Imprisoned	Budapest residents*
I	1.5	2.2
II	2.8	5.3
III	4.4	5.6
IV	3.4	3.9
V	2.8	2.7
VI	4.5	3.7
VII	7.4	4.9
VIII	7.8	5.4
IX	6.7	4.4
X	6.0	4.6
XI	4.7	8.4
XII	1.8	4.0
XIII	8.7	6.8
XIV	6.4	8.1
XV	5.5	5.4
XVI	3.0	3.5
XVII	2.8	2.7
XVIII	4.1	4.5
XIX	3.2	3.0
XX	6.1	5.0
XXI	3.5	3.5
XXII	2.9	2.4
Total	100.0	100.0

Note: * for location of Budapest districts see Figure 5.1

distributions are the same – the value for the sample of imprisoned offenders was 20.4 when compared to the population over 14 years old. This value is higher than the segregational index for unskilled and semi-skilled workers (17.0) but lower than the index for the gypsy population (53.9). Thus we can conclude that the extent of the spatial segregation of offenders is similar to the spatial segregation of groups of low occupational status but not as extreme as it is for the gypsy population.

These conclusions are reinforced by looking at the residence patterns of semi- and unskilled workers, gypsies and imprisoned offenders (Figures 5.1, 5.2 and 5.3). As has already been suggested, the spatial segregation of different occupational groups is characterized by the contrast between the large concentration of high-status groups in parts of Buda and the dispersed but clearly definable small areas of low-status groups on the Pest-side of the capital. The area where gypsies live in the slums of inner Pest,

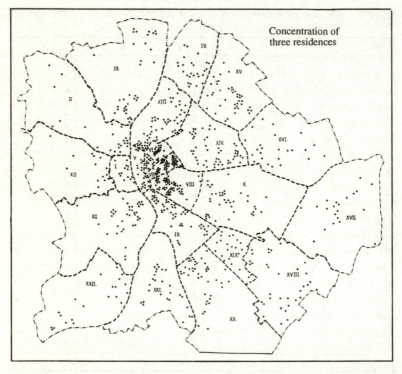

Figure 5.3   Residential distribution of those in prison in Budapest 1980

however, forms a large, coherent area. The spatial distribution of imprisoned offenders is more even, except that they tend to be over-represented in the inner, slum areas of Pest and under-represented in the high-status parts of Buda.

Next the relationship between the type of offender and the kind of area in which they lived was considered. To do this the 490 census tracts are aggregated into eleven clusters on the basis of their socio-economic characteristics (see Table 5.4). As can be seen from Table 5.4 the status of a cluster is closely associated with the proportion of manual workers among the economically active population.

Table 5.5 shows that there is a relatively strong positive correlation between imprisoned offenders in a cluster and the proportion of manual workers in that cluster. More particularly, there is also a clear relationship between the type of offender and the socio-economic status of a cluster. Recidivists and offenders with a low educational level are more common in low-status clusters than offenders with a higher educational level or who have committed their first offence. This relationship can be expressed numerically in terms of correlation coefficients. The location of those imprisoned offenders who are relatively educated or who are in prison for the first time shows a very low correlation with the proportion of semi-skilled and unskilled workers by cluster (0.02) while the correlation coefficient for the uneducated recidivists is relatively high (0.51). This suggests that the latter type of offender is concentrated in the lowest-status parts of the city.

Figure 5.4 shows the proportion of offenders by cluster in Budapest. The map shows that on the periphery of the city in agricultural and industrial areas, where there are clusters with very small populations, there is a high concentration of offenders. But if Figure 5.4 is compared with Figure 5.3 it is apparent that we are dealing with very small localized concentrations of imprisoned offenders, not large areas. Although the scale of the spatial units in Figure 5.4 is quite coarse it is still possible to distinguish between the higher-status areas, the city centre and the low-status areas. For example, one can see a large spatially coherent area in Buda where the proportion of people who have been imprisoned is very low. This area is characterized by private apartments and has always been the high-status part of the city. Thus the area with the lowest proportion of detected and convicted criminals coincides almost exactly with the highest status parts of the city.

Table 5.4 Distribution of social–occupational groups by cluster

Cluster number	Cadre elite	Professionals	Other white-collar workers	Production managers	Skilled workers	Semi-skilled and unskilled workers	Not employed in state sector
1	8.9	15.7	18.5	1.6	8.9	7.1	1.0
2	6.0	1.5	13.3	7.6	30.5	18.9	0.1
3	4.0	4.7	20.8	3.0	23.0	21.5	0.7
4	5.5	8.1	18.1	2.0	13.0	12.4	1.2
5	4.3	5.6	16.5	2.1	15.5	15.3	1.6
6	3.2	3.6	15.3	2.1	17.4	18.4	1.1
7	7.8	9.4	23.6	3.0	17.2	13.4	0.8
8	14.7	17.8	23.3	2.2	9.1	6.4	1.2
9	6.8	11.1	18.6	1.8	10.7	9.9	1.2
10	4.4	5.6	17.6	2.3	16.2	14.7	0.7
11	0.0	5.3	9.8	1.5	15.2	19.7	15.2
12	2.9	3.0	11.6	2.9	17.8	28.2	3.8
13	2.5	2.2	13.8	2.0	21.3	24.3	0.9
14	3.5	0.0	15.4	0.0	19.2	8.8	0.0
15	1.2	1.0	6.9	0.8	20.0	48.2	0.3
Average	4.9	6.2	17.0	2.2	16.8	17.8	1.2

| Cluster | Retired | | | Other | Total | Cadre elite and professionals and retired college educated | Manual worker per active earner | Semi-skilled and un-skilled worker per active earner |
	College educated	Secondary school degree	Eight grades or fewer of primary school education					
1	6.2	9.5	20.1	2.5	100 (N=109,279)	30.8	28.5	11.5
2	0.3	1.3	18.5	1.8	100 (N=219)	7.8	73.1	24.2
3	0.4	1.8	16.0	3.9	100 (N=179,847)	9.1	61.1	27.7
4	2.9	6.7	27.5	2.5	100 (N=142,877)	16.5	45.4	20.6
5	1.5	4.6	30.1	3.0	100 (N=204,703)	11.4	54.0	25.1
6	0.8	2.9	32.2	3.1	100 (N=241,445)	7.6	62.0	30.1
7	1.3	3.2	17.6	2.6	100 (N=159,221)	18.5	44.7	17.8
8	3.4	4.8	14.2	3.0	100 (N=55,864)	35.9	23.7	6.6
9	4.6	8.2	24.4	2.4	100 (N=133,355)	22.5	37.3	16.5
10	1.5	4.5	29.9	2.5	100 (N=120,868)	11.5	54.0	23.9
11	0.0	2.3	23.5	7.6	100 (N=55)	5.3	54.6	29.5
12	0.2	2.9	22.0	4.7	100 (N=4,745)	6.1	69.6	40.3
13	0.3	1.6	27.3	3.6	100 (N=124,565)	5.0	71.0	36.3
14	0.0	0.9	52.3	0.0	100 (N=35)	3.5	59.7	18.8
15	0.1	0.8	15.5	5.2	100 (N=2,540)	2.3	88.0	61.5
	1.7	4.0	25.2	3.1	100 (N=1,479,618)	55.7	26.9	27.0

Table 5.5 Proportion of those who served their sentence in 1979 (in the over 14 years age group) and the linear correlation coefficients by cluster (%)

Cluster	Number of population	First imprisonment <8 grades of elementary school	First imprisonment >8 grades of elementary school	Recidivists <8 grades of elementary school	Recidivists >8 grades of elementary school	First imprisonment total	Recidivists Total	Fewer than 8 grades of elementary school	More than 8 grades of elementary school	Total
8	61,525	0.07	0.24	0.11	0.10	0.31	0.21	0.18	0.34	0.52
1	120,126	0.17	0.16	0.17	0.14	0.33	0.31	0.34	0.30	0.65
9	145,280	0.19	0.34	0.42	0.29	0.53	0.71	0.61	0.63	1.25
7	172,375	0.21	0.18	0.36	0.17	0.39	0.53	0.57	0.35	0.94
4	154,810	0.33	0.30	0.59	0.28	0.63	0.88	0.92	0.58	1.52
10	130,086	0.30	0.34	0.67	0.32	0.64	0.98	0.97	0.65	1.64
5	221,528	0.38	0.27	0.73	0.40	0.65	1.13	1.11	0.67	1.80
3	194,020	0.29	0.23	0.38	0.13	0.52	0.51	0.67	0.36	1.04
6	259,989	0.39	0.36	0.75	0.32	0.75	1.06	1.14	0.67	1.85
12	5,201	1.35	0.19	1.54	0.58	1.54	2.11	2.88	0.77	4.23
13	134,716	0.48	0.23	0.82	0.33	0.71	1.15	1.31	0.56	1.91
15	3,400	0.59	0.29	4.41	0.59	0.88	5.00	5.00	0.88	5.88
Total	1,603,056	0.31	0.27	0.56	0.26	0.58	0.82	0.87	0.53	1.42
Correlation coefficients		0.38	0.02	0.51	0.24	0.30	0.46	0.53	0.14	0.43

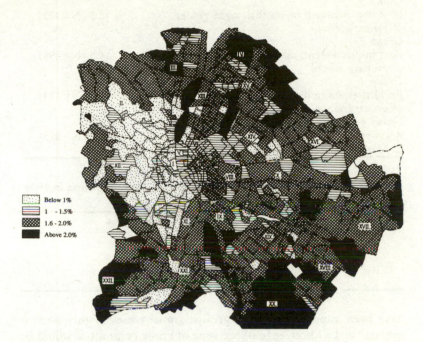

Figure 5.4 Residential distribution of those in prison in Budapest in 1979
by cluster

The spatial distribution of imprisoned offenders can also be
examined by using a segregational index (Table 5.6). This analysis of
segregational indices further reinforces an earlier conclusion that the
spatial segregation of imprisoned offenders is a function of the
social status of these groups. The dissimilarity indices in Tables 5.7
show the reciprocal spatial segregation of the above groups. As we
have already observed the segregational indices of relatively uneducated first offenders and relatively educated recidivists are very
similar. On the basis of the dissimilarity indices it seems that
Groups 2 and 3 are more segregated from each other on the basis of
their educational level than on the number of times they have been
imprisoned.

So far we have examined four groups of imprisoned offenders on
the basis of their educational level and the number of times they

Table 5.6 Segregational indices of those in prison in 1979 compared to the population (over the age of 14 years) by cluster

Group 1	
First imprisonment, more than eight grades of primary school	10.5 (N=435)
Group 2	
First imprisonment, maximum eight grades of primary school	14.5 (N=496)
Group 3	
Recidivists, more than eight grades of primary school	16.1 (N=424)
Group 4	
Recidivists, maximum eight grades of primary school	18.2 (N=895)

Table 5.7 Dissimilarity indices of some of the groups in prison in 1979 by cluster

	Dissimilarity Index
Groups 1–2	12.2
Groups 1–3	11.4
Groups 4–3	8.6
Groups 4–2	6.8

have been imprisoned. These groups are of course highly heterogeneous and a knowledge of the type of crime committed would be required to form more homogeneous groups. Unfortunately these data are not available in terms of residential location of the imprisoned offenders. However, some information can be gained less directly by examining the length of sentences of imprisoned offenders which is clearly associated with the type of crime (Table 5.8). From Table 5.8 it can be seen that the segregational indices increase in line with the number of years spent in prison. This can be explained partly by the different number of cases but is also linked to the fact that the proportion of low-status people with a low level of education is higher among those with longer prison sentences. This can be partly attributed to the fact that underprivileged people are often sentenced for longer than more privileged people who have committed the same crime, because of their inexperience and lack of proper legal defence.

Table 5.8 Segregational indices of those in prison in 1979 by number of years in prison

	In prison for more than three years		In prison for more than four years		Total	
	Seg. Index	N	Seg. Index	N	Seg. Index	N
Fewer than eight grades in primary school	19.4	485	21.1	293	16.8	1391
More than eight grades in primary school	14.5	232	18.3	137	12.4	859
First imprisonment	16.8	259	18.1	163	10.6	931
Recidivist	20.0	458	20.5	267	17.3	1319
Average	17.4		18.4		14.6	

CONCLUSION

The analysis of the data on imprisoned offenders in Budapest contained in this chapter does not support the view that there is a mass criminal subculture living in a certain part or parts of the city. Furthermore, in those parts of the city which are portrayed as 'criminal areas' in police reports, such as the inner city slum areas of Pest, the proportion of imprisoned people is usually no higher than the proportion one would expect given the low status of the people living in these areas.

Acknowledgement

The English translation of this chapter was prepared by Dr Julia Meszaros.

REFERENCES

Andorka, R. (1982) *A Tarsadalmi mobilitás valtozasai Magyarorszagon* (Changes of social mobility in Hungary), Budapest: Gondolat Kiado.

Duncan, O.D. and Duncan, B. (1955) 'Residential distribution and occupational stratification', *American Journal of Sociology* 493–503.

Ferge, Z. (1969) *Tarsadalmunk retegezodese* (Stratification in our society), Budapest: KJK.

Kolosi, T. (1984) *Statusz es reteg* (Status and strata), Budapest: Tarsadalomtudomanyi Intezet Kiadvanyai.

Ladányi, J. (1988) 'A lakohelyi szegregacio alakulasa Budapesten 1930 es 1980 kozott' (Residential segregation in Budapest between 1930 and 1980), *Valosag* 3.

Ladányi, J. (1989) 'Changing patterns of residential segregation in Budapest', *International Journal of Urban and Regional Research* 13 (14).

6

THE JOURNEY TO CRIME
Conceptual foundations and policy implications
George F. Rengert

Journey-to-crime studies have their conceptual basis in earlier journey-to-work studies. The basic assumption is that, like work, crime is a primary source of livelihood for the professional criminal. Like workers, most criminals must travel from their homes to a crime site to make a living. In certain cases, professional criminals become professional because crime and work are mutually exclusive due to the accessibility constraints of conflicting space and time demands (Rengert and Wasilchick 1985). Criminals cannot be at a work site and a crime site at the same time.

A complementary view of the journey to crime is to conceptualize it as a transportation problem. In the case of property crime, commodities which are stolen (whether money or goods) must be transported back to the home of the criminal, to a storage location, or to a sales location (fence). In this case, concepts developed in transportation research have possible application to the analysis of the journey to crime.

Following the transportation perspective, the journey to crime can be conceptualized as three distinct phases. First is the origin point (usually home) and the directional orientation the criminal travels with respect to an anchor point. This phase often involves making one's way to a major traffic artery and proceeding in a determined direction where the second phase of the journey begins. The second phase is that of traversing (or bridging) space. The objective is to locate a search space by passing over areas which are not considered useful for criminal purposes. When a useful community or region is identified, the third phase of the journey begins. This is the active search phase which continues until a likely site is identified.

As yet, we do not have a study which integrates all three phases of the journey to crime into a single analysis from the transportation perspective. Past research has focused on only one or another part of this conceptualization. As a result, we do not have an integrated view of the spatial aspects of the journey to crime. In the following sections, each phase of the journey to crime is discussed and important research focused on each is identified. Public policy associated with each phase is discussed and evaluated. In the final section, the integration of these studies into a unified whole is discussed.

ANCHOR POINTS AND DIRECTIONAL ORIENTATION

Early studies of directional orientation were methodological investigations which did not address the reasons for the directional bias. Rather, the purpose was to establish that directional bias existed and that criminals did not travel in random directions in their journeys to crime sites (Costanzo *et al.* 1986; Lenz 1986). One study did suggest that the central business district is a primary anchor point for property criminals in a large metropolitan region (Rengert 1989). These were aggregate studies of large urban regions.

Specific anchor points for individual criminals can be identified only when we break down these aggregations and examine the spatial behaviour of individual criminals. This is because anchor points are often specific to the individual criminal. For example, Rengert and Wasilchick (1985) established that work places tended to orient the direction of crime search for residential burglars who were employed within six months of their arrest. For those who were not employed, recreation sites oriented the direction of the search for a crime site (Rengert and Wasilchick 1985). Since work and recreation sites varied with each individual criminal, an aggregate analysis would not have identified a directional bias. Individual level analysis is required.

These studies are more than simple intellectual exercises. Important public policy issues are related both to the aggregate and to the individual level studies. The aggregate level analyses established that any public policy which increases the level of criminality contributes to 'spatial injustice' (Rengert 1989) since all residents of the region do not experience an equal increase in victimization. For example, early release programmes to relieve prison overcrowding lead to an increase in 'avertible recidivism' (Greenfeld 1985).

110

Avertible recidivism is crime committed by the early releasees which would not have been committed had they served their entire sentences. If all residents of the region share equally in the increased victimization associated with avertible recidivism, it might be a justifiable policy. Residents may choose to accept the increased level of crime rather than pay the costs of incarceration or community corrections. However, if the increased crime is focused on a few (usually inner city) communities who are not as politically powerful as the wealthier suburban residents, then issues of spatial injustice are called into question. Low-income inner city communities serve as dumping grounds (or surrogate prisons) whenever the criminal justice system becomes overburdened (Rengert 1989). Directional studies of the journey to crime at the aggregate level are necessary to determine if all members of society share equally in any increase in criminality associated with decisions made by criminal justice professionals.

At the individual level of analysis, directional analysis of the journey to crime also has important public policy implications. Recent research by Rengert and Wasilchick (1989) suggests that drug sales locations serve as important anchor points for drug-dependent property criminals who finance their drug habits through property crime. The crime sites of these drug-dependent property criminals tend to be clustered about the drug sales locations rather than about the home of the criminal as is the general case (Brantingham and Brantingham 1984). The exact reason for this directional bias toward drug sales locations is yet to be determined. Rengert and Wasilchick (1989) suggest that the criminal attempts to minimize distance by committing the crime on the way to a drug purchase. However, Cromwell *et al.* (1989) suggest that drug offenders may purchase a 'fix' before they attempt their first property crime of the day. Their research suggests that most drug-dependent offenders use heroin or marijuana to get in the correct frame of mind to commit crimes. Earlier research by Rengert and Wasilchick (1985) also found criminals who use drugs to relax them or, in one case, to enhance his sense of hearing. Many drug-dependent property criminals feel a need for drugs to prepare them for a crime. If they purchase the drug before searching for a crime site, their spatial search may begin from the drug purchase area rather than from the home.

In either case, important policy issues are related to the impact of this spatial concentration of crime on the communities which surround a drug sales location. Open street sales of illegal drugs are possible only in communities where the social fabric is so disrupted that effective community resistance to drug sales is not possible. Police reason that they cannot operate effectively without community co-operation (Wilson and Kelling 1982). Therefore, police may practise a containment policy of largely ignoring the social and criminal problems of these drug sales regions. However, this containment policy is doomed to failure due to the spatial relationship of property crime to the drug sales location.

If drug-dependent property offenders focus their crime on communities surrounding the drug sales location, the remaining social fabric of these surrounding communities will be destroyed by the spatial concentration of crime as well. This allows drug sellers to expand operations into the surrounding communities. Property crime will probe further outward. There lies the spatial dynamic of expanding drug sales. Drug sellers and drug-dependent property criminals may operate in a symbiotic relationship spatially. If the police ignore the heart of the problem through a containment policy, the problem will grow spatially until whole sections of the city are impacted. It will not stay static spatially. To be effective, police and community action workers need to operate without community co-operation in high crime drug sales regions. Further research to document this symbiotic relationship is needed. If it is verified, containment policies must be abandoned if we are to deal effectively with the drug–crime relationship.

BRIDGING SPACE

The second phase of the journey to crime is that of bridging the territory which is not considered appropriate for criminal activity. Most studies which focus on this phase are a variation of the common distance to crime studies. Early works compared the distance travelled by criminals from their residences to crime sites for different types of crimes and for different classes of criminals (Pyle 1974). Little attention was placed on reasons for the differences in distances to crime.

Turner (1969) analysed the distances delinquents are likely to travel from their home to commit a crime. He identified a block or two around the home that was avoided by the delinquents. Crimes peaked just beyond this zone of avoidance. Turner reasoned that delinquents avoid the blocks immediately around their home because of the risk of recognition in familiar neighbourhoods. He was one of the first to recognize that criminals not only attempt to minimize distance to crime, but also minimize recognition by avoiding their own residential neighbourhoods.

Hakim and Weinblatt (1984) adopted the familiar Von Thunen land use model to explain the distances criminals engaged in various types of property crimes are likely to travel to commit their crime. The original Von Thunen model is used to explain the spatial arrangement of agricultural production around a farming village. The idea is, the bulkier the product produced, the closer to the village it will be produced. Hakim and Weinblatt (1984) argued that criminals harvest an area with constraints in common with farmers. They argued that the bulkier the commodities stolen, the closer to home it will be stolen. They demonstrate that residential burglars who steal bulkier items than robbers (who generally steal money) victimize areas closer to their homes.

Rengert and Bost (1978) relaxed the linear assumptions of Hakim and Weinblatt's model in order to account for the avoidance of recognition in their own neighbourhoods for criminals who commit confrontational crimes. They assume that a criminal will attempt to commit their crime in as familiar territory as possible while at the same time attempting to avoid recognition in confrontational crimes such as robbery. Therefore, the expectation is that a mathematical curve fit to the distribution of crimes committed in distance zones outward from the residences of criminals will be concave for residential burglars since there is no confrontation in this sneak type of crime. A convex curve is expected for armed robbery where confrontation exists. When these crime zones are rotated 360 degrees around the home of criminals, high crime zones are identified for each type of crime much like production zones are identified for crops around an agricultural village.

The Rengert and Bost (1978) study has important policy implications. It was designed to measure the areal extent of the impact of the public policy decision to house low-income subsidized housing residents in large aggregated units. Not only is the size of the aggregation important, but also in what section of the city it is placed. In North American cities, there is a tendency to segregate the poor in inner cities. Therefore, the non-criminal poor bear the brunt of a policy which segregates even more poor in their communities rather than a policy of scattered site housing, vest pocket projects, or subsidized rents.

SEARCH OF SPACE

The search of space phase of the criminal journey to crime begins when criminals actively start to evaluate their surroundings in order to locate a crime site. It begins at the termination of the bridging space part of the journey to crime and ends when a suitable crime site has been chosen subjectively. The extent of the search phase has been examined in a variety of ways. The temporal extent of the active search has been evaluated by Rengert and Wasilchick (1989) using various 'stopping rules' to determine when the search begins and ends. It was discovered that the extent of the search depended on the urgency of the need of a residential burglar for money. For example, drug-dependent burglars who had not managed their drug supply adequately would burgle the first empty house they encountered. Both the bridging of space and the active search phase of the journey to crime were truncated.

At the other extreme, residential burglars who enjoyed the search process (often making a game of picking just the right house) had extensive searches in which they developed a short list from which they picked the best house. They would then retrace their journey back to this house to commit a burglary. This type of burglar is best described by the 'house hunting' stopping rule (Flowerdew 1976) where many houses are evaluated before one is decided upon.

Between these two extremes is a type of residential burglar who has a previously established aspiration level which he tries to reach in the spatial search process. He searches until he finds a house which is above this aspiration level which he then burgles. This type of burglar is described by the 'Marriage problem' stopping rule (Flowerdew 1976). These stopping rules are used to describe in general terms the scenario through which the active search for a crime site terminates and the commission of the crime usually begins.

The characteristics of specific sites that are chosen for crime have been examined by Bennett and Wright (1984) and Zahn (1989). Both studies focused on the environmental attributes of sites. Bennett and Wright examined characteristics of homes which residential burglars (acting as expert witnesses) identified as positive and negative attributes. Zahn identified characteristics of convenience stores which experienced a disproportionate number of robberies. Sherman et al. (1989) have examined the characteristics of sites which placed a disproportionate number of calls for service to a local police department. These places are termed 'hot spots'.

Again, there are important public policy implications associated with these studies. Individual home-owners can alter the environmental characteristics of their homes. One such alteration might be the addition of a burglar alarm. However, this begs the question of whether alarm systems deter criminals or merely displace crime spatially on to unalarmed homes. If burglar alarms reduce the number of burglaries, then alarm installations should be encouraged. Police save in crime investigation costs as well as residents saving in crime costs. If alarms displace burglaries spatially, then there are serious questions concerning their usefulness, especially if they contribute to spatial injustice by focusing crime on households and communities which may not be able to afford the high cost of burglar alarms. In this later case, income transfer is taking place where unalarmed homes are bearing a disproportionate share of the cost of burglary. Furthermore, alarmed homes receive greater police protection since police answer false as well as legitimate activations with one or two cars. The policy question is whether public officials should encourage the installation of burglar alarms or discourage them with high initial fees and escalating charges for false activations.

Hot spots which require much more attention from the police also need to be addressed from a policy perspective. If these hot spots are a public noxious site, we need to consider whether they should be destroyed like crack houses which sell drugs in many North American cities. On the other hand, if they are legitimate business establishments, should they be made to pay for their overuse of police services in comparison to other business establishments. Especially disorderly bars may be held accountable for some of the problems they generate. Retail establishments may be required to hire private security to supplement the public police. These issues may require a case by case consideration to determine the degree of private responsibility which must be assumed by the local hot spots.

THE JOURNEY TO CRIME RECONSIDERED

The journey to crime contains several distinct yet integral phases which can be considered when public response to the crime problem is considered. Public officials must be aware not only of where criminals commit their crimes, but also of where the criminals originate and the spatial impact of policies designed to lower crime at specific sites and in specific communities. At the heart of this issue is the spatial displacement of crime and the official responsibility for displacement. Public officials must consider not only who benefits from any policy, but also who may pay a higher price as crime is shifted from place to place. There are no easy answers. Additional information will lead to a fuller understanding of the spatial impacts of our policies. We require as complete an understanding of the spatial dynamics of our policies as possible if we are to approach fair and just cities in our societies.

REFERENCES

Bennett, T. and Wright, R. (1984) *Burglars on Burglary: Prevention and the Offender*, Aldershot: Gower.

Brantingham, P. and Brantingham, P. (1984) *Patterns in Crime*, New York: Macmillan.

Costanzo, C., Halperin, W. and Gale, N. (1986) 'Criminal mobility and the directional component in journeys to crime', in R. Figlio, S. Hakim and G. Rengert (eds) *Metropolitan Crime Patterns*, Monsey, NY: Criminal Justice Press.

Cromwell, P., Olson, J. and Avary, D. (1989) *Residential Burglary: A Staged Activity Analysis*, Final Report to the National Institute of Justice (Grant 8-7602-TX-IJ), Washington, DC: US Department of Justice.

Flowerdew, R. (1976) 'Search strategies and stopping rules in residential mobility', *Transactions: Institute of British Geographers* ns 1: 47–57.

Greenfeld, L. (1985) *Explaining Recidivism*, Washington, DC: US Department of Justice, Bureau of Justice Statistics.

Hakim, S. and Weinblatt, J. (1984) 'The impact of criminal mobility on land prices: a theoretical view', *International Journal of Social Economics* 11: 24–30.

Lenz, R. (1986) 'Geographical and temporal changes among robberies in Milwaukee', in R. Figlio, S. Hakim and G. Rengert (eds) *Metropolitan Crime Patterns*, Monsey, NY: Criminal Justice Press.

Pyle, G. (1974) *The Spatial Dynamics of Crime*, Department of Geography Research Paper 159, Chicago: University of Chicago Press.

Rengert, G. (1989) 'Spatial justice and criminal victimization', *Justice Quarterly* 6 (4): 543–64.

Rengert, G. and Bost, R. (1978) 'The spillover of crime from a housing project', Paper presented to the Academy of Criminal Justice Sciences, St Louis, Missouri.

Rengert, G. and Wasilchick, J. (1985) *Suburban Burglary: A Time and a Place for Everything*, Springfield, Ill: Charles Thomas.

—— (1989) *Space, Time, and Crime: Ethnographic Insights into Residential Burglary*, Final Report to the National Institute of Justice (Grant 88-IJ-CX-0013), Washington, DC: US Department of Justice.

Sherman, L., Gartin, P. and Buerger, M. (1989) 'Hot spots and predatory crime: routine activities and the criminology of place', *Criminology* 27 (1): 27–55.

Turner, S. (1969) 'Delinquency and distance', in T. Sellin and M. Wolfgang (eds) *Delinquency: Selected Studies*, New York: Wiley.

Wilson, J.Q. and Kelling, G.L. (1982) 'Broken windows: the police and neighborhood safety', *Atlantic Monthly* 255: 29–38.

Zahn, D. (1989) 'The crime prevention effort in the state of Florida', Paper presented to the American Society of Criminology, Reno, Nevada.

7

HOUSING MARKETS AND RESIDENTIAL COMMUNITY CRIME CAREERS
A case study from Sheffield

Anthony E. Bottoms, Ann Claytor and Paul Wiles

In 1986 we published a paper which contained as its central argument the claim that 'the key to an understanding of [offender-based] residential community crime careers lies in the operations of the housing market' (Bottoms and Wiles 1986: 103). In this chapter we propose to elaborate that claim with particular reference to empirical work carried out since the 1986 paper was written.

An important preliminary task is to clarify the meaning of the three main concepts used within the central theoretical claim. First, the concept 'residential community crime career' derives from an observation of Albert Reiss (1986) that 'communities, like individuals, can have careers in crime, [and] today's safe environment can become tomorrow's dangerous one'. Use of the concept implies that one will attempt to understand the ways in which the social life of particular communities might change over time, and how these changes interact with changing crime and offender patterns. Second, the theoretical claim speaks of offender-based residential community crime careers. This picks up the distinction, familiar to environmental criminologists, between area offence rates and area offender rates: 'area offence rates' refer to the extent to which offences are committed in particular localities, while 'area offender rates' refer to the extent to which known offenders live in particular areas. The distinction between the two rates is often of great importance, and it has been known to criminologists for many years (see Morris 1957), yet it is still not sufficiently appreciated by many. A residential community crime career can be understood either in offence-based terms or in offender-based terms, but our central claim refers especially to offender-based residential community crime careers (but see p. 122). The third concept in our central claim is that of the 'housing market'. Elsewhere we have defined this concept in wide terms as referring to 'all processes which enable people to move into residential properties (by buying or renting), or which inhibit them from doing so; and also to all processes enabling or inhibiting the termination of residence in a property when desired' (Bottoms and Wiles 1988: 84). A broad definition of this kind, of course, covers a much wider ambit than simply the economics of private house-buying, and it includes among other things the bureaucratically determined rules of public housing authorities in Britain.

So much then for preliminary conceptual clarification. Turning to the substance of our central theoretical claim, how can it be argued that the housing of an area affects its offender or offence rates over time? One relevant consideration here is the possible influence of housing and area design. Design is an important aspect of residential areas, the criminological significance of which has been much stressed by some writers, for example Newman (1972) and Coleman (1985), the latter particularly in dispositional (or offender-based) terms (see Coleman 1989). We do not seek to deny the importance of design for residential community crime careers in certain circumstances, although we would not place as much stress upon it as Coleman; rather we agree with Taylor and Gottfredson's (1986: 411) assessment of the evidence: 'it appears that alteration of physical environment features cannot have stand-alone crime prevention effectiveness. Resident dynamics are key mediators of the environment–crime linkage.' In agreeing with this statement, we think it is also worth pointing out that the phrase 'resident dynamics' has a usefully wide meaning, embracing as it does both the movement of families in and out of areas, and the social life of communities within areas. It is precisely these elements of 'resident dynamics' which, in our view, are related both to the housing market and to offending in particular areas.

It is quite important to spell out in a little more detail exactly what we mean by our central claim. One obvious aspect of the claim is that the allocation to different areas of social groups with different offending propensities can in itself have a criminogenic or non-criminogenic effect: for example, if the housing market results (directly or indirectly) in a particular area being populated

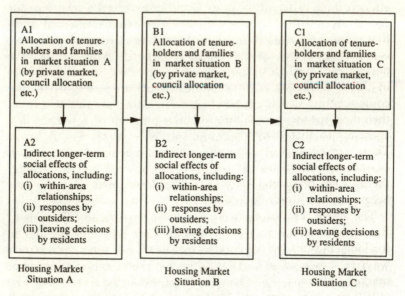

A1 Allocation of tenure-holders and families in market situation A (by private market, council allocation etc.)	B1 Allocation of tenure-holders and families in market situation B (by private market, council allocation etc.)	C1 Allocation of tenure-holders and families in market situation C (by private market, council allocation etc.)
A2 Indirect longer-term social effects of allocations, including: (i) within-area relationships; (ii) responses by outsiders; (iii) leaving decisions by residents	B2 Indirect longer-term social effects of allocations, including: (i) within-area relationships; (ii) responses by outsiders; (iii) leaving decisions by residents	C2 Indirect longer-term social effects of allocations, including: (i) within-area relationships; (ii) responses by outsiders; (iii) leaving decisions by residents
Housing Market Situation A	Housing Market Situation B	Housing Market Situation C

Figure 7.1 Diagrammatic representation of the relationship between the potential effects of the housing market and residential community crime careers

exclusively by people over 70, the offender rate of that area will be extremely low. Some commentators have assumed that our claim about the relationship of the housing market to offender-based residential community crime careers amounts to no more than such direct allocative effects. It is therefore vital to emphasize that this is not the case. Rather our claim is that the relationship of the housing market to offender-based residential community crime careers can be understood as a process with three elements (illustrated in Figure 7.1), only one element of which refers to direct allocative effects.

Turning then to the other features of Figure 7.1, within any given housing market context one has to consider the indirect and longer-term social effects of initial allocations. By this we mean the interactive effects of residents with their physical environment, and the social processes that occur between and within the social groups who find themselves living in that particular area as a result of choices made within the housing market's allocative mechanisms. These indirect effects include the formation of political associations and voluntary groups in the area, as well as the development of cultural styles and patterns. Our argument is that these indirect effects can, in principle, be more important in shaping a community crime career than the initial propensity to offend indicated in the first stage of the model. Yet the specific form of the indirect effects is in part dependent upon the initial allocative process: residents can develop the social life of their areas only after they have come to live in them. Additionally, of course, once a group of people have come together in an area and begun to develop a community life, then the response by outsiders (including social control and welfare agencies) to that community is potentially of great importance. These various socio-cultural developments, including the reaction of outsiders, can also affect the leaving patterns of persons living in the area, so there may be a greater or lesser wish to leave areas because of the kind of social life that has been developed.

The above discussion assumes a static housing market context. But, of course, in reality the housing market context is never static for long, and this introduces the third element in Figure 7.1, where we have shown, in an oversimplified way, three successive housing market situations. As originally emphasized by Rex and Moore (1967) in their work on 'housing classes', and as subsequently developed by the 'urban managerial' school of the 1970s, changes in the local housing market can substantially affect the composition of an area's population, and consequently the social life of areas, in a number of ways:

1 Within a private housing market, differential house price movements in different areas can radically affect the resident dynamics of areas (Taub et al. 1984).
2 Within the public housing sector, changes in the local authority's allocation rules can have profound longer-term effects on areas, as can, for example, the building of a new council estate with good amenities next to an old and somewhat dilapidated housing area.
3 The reputations of areas can change, for reasons either connected with or unconnected with other housing market changes; but once a reputation changes, for whatever reason, this can have a powerful effect both on residents' decisions to leave an area and on the kind of potential residents who may consider moving in to the area.

In Figure 7.1 we have shown just three housing market situations, but of course in reality the housing market (in its broad definition, as discussed at the beginning of this chapter) is always changing. It changes partly because of factors external to particular local areas (the general economic situation in the country, the interest rate level set by a central bank, government control on public sector borrowing, and so on), but also partly because of much more local variables, including the way in which the social life of a particular area develops, and is viewed by outsiders (that is, the second feature in Figure 7.1). Thus Figure 7.1 shows, in outline form, what is in reality a continuous dynamic three-feature interactive situation.

Reduced to its bare essentials, then, our central claim, based on the three-feature interactive model outlined in Figure 7.1, is that in order to understand and explain offending behaviour by residents of particular areas, it is vital to consider who lives in these areas; how they came to live there in the first place; what kind of social life the residents have created; how outsiders (including official agencies) react to them; and why they remain in the areas and have not moved. The model provides a general framework for analysis, but is not itself explanatory or predictive until applied to the empirical details of a particular area.

While it should by now be obvious how all this is related to offender-based residential community crime careers, the link of the housing market to offence-based residential community crime careers is less clear cut, and most straightforwardly depends on the extent to which crime in an area is committed by local people. Commonly area offences are committed by local people, but if in a given area most offences are committed by outsiders, then the link between the housing market and residential community crime careers will not operate so clearly. Yet even offence-based residential community crime careers involving non-resident offenders can be related to the housing market. For example, one of the areas examined in the Sheffield study (but not discussed in detail in this chapter) was a classic interstitial area. However, it was not just the inner-city environment of the area which was important, but the fact that it was one of the few districts which fulfilled the function of providing multi-occupation, short-term privately rented accommodation in the city's housing market. The result was a social milieu within which the supply of illicit goods and services flourished. The resulting high area offence rate was produced not only by offenders living in the district (although the area did in fact have a fairly high offender rate), but also by offenders being attracted from outside the area by the deviant opportunities it offered.

Before leaving this initial theoretical section, we want to make it abundantly clear that the housing market does not operate alone in the creation of offender-based residential community crime careers. Rather, it interacts with a range of other aspects of social life to create the relevant social effects. These other social aspects include:

1 the *social networks* of the area
2 *socialization processes* affecting the children and youth in the area (including family life, peer groups, and schools: see generally Martens 1990; Lindström 1990)
3 the work of *social control agencies* in the area (see Gill 1977)
4 the development of *reputations and labels* about the area and its residents, and the effects of these labels both upon residents themselves and upon potential residents (labelling and stigma theory)
5 the *economic development* of the area in the context of the city as a whole (for the interaction of this with housing market decisions and other social procëses see Taub *et al.* 1984)
6 the *physical form* of the area, including both natural features and the design of the built environment, and the way these features interact with the social life of the area.

We shall return to the significance of some of these matters later in the chapter. For the moment it is sufficient to note two features of our approach. First, as the above list makes clear, the approach is an eclectic one: however, as the work of Braithwaite (1989) makes clear there is nothing wrong with eclecticism as long as the various matters being considered are related to one another in a coherent way. (Indeed, since social life is experienced by people *as a whole* eclecticism certainly has more to offer than a single-factor social theory which is blind to many features of social life.) Second, the approach emphasizes the importance of place, and the dynamics of small-scale communities, in a way which has often been neglected in post-war social theory, but is now being rediscovered (e.g. Giddens 1984; Gregory and Urry 1985): we discuss this matter more theoretically elsewhere in this volume (see Chapter 1).

THE SHEFFIELD STUDY

In the remainder of this chapter we propose to look at some empirical work carried out in Sheffield in which we have tried to understand the residential community crime careers of certain areas.

That empirical work, we believe, illustrates and elaborates the central theoretical claims we have just made. Our focus here will be on just two areas, but before we describe them it is necessary to give a little background about the Sheffield study of urban crime as a whole.

The first stage of the Sheffield study consisted of a general analysis of recorded crime in the whole of the city (Baldwin and Bottoms 1976). Among the results of that study was the apparent importance of tenure-type in the general statistical analyses of offender rates in different parts of Sheffield, plus some hints of the apparent importance of the allocative processes within the large council housing sector in the city. These findings led to a greater emphasis on the housing market in our subsequent work.

All subsequent stages of the Sheffield research have concentrated on just a few small areas in the city. Six areas in particular have been studied, and these six areas have been analysed both in the mid-1970s and, more recently, in a replication study carried out in 1988. The six areas fall into three groups: one pair of pre-war low-rise council estates, one pair of post-war high-rise council estates, and one pair of low-rise housing areas in the private sector. When we first examined these areas in the 1970s, within each pair one of the areas had a high official offender rate and the other a low offender rate; in those areas at that time there was also a correlation between the official offender rate and the official offence rate for the area, so that the areas also tended to have a high or a low offence rate as well. Moreover, within each of the pairs police-recorded offence rates correlated well with victim survey offence rates (for a much fuller account of these results see Bottoms *et al.* 1987).

For reasons of space, our focus in this chapter is on only the first of these three pairs, that is the pre-war council pair, which we shall call Gardenia and Stonewall. Gardenia and Stonewall were both built in the first quarter of the twentieth century, and they both belonged to that early phase of council house building in Britain which focused on the artisan sector of the working class and was designed to provide a good environment of a 'garden city' type for the skilled worker and his family (see Merrett 1979). They are both low-rise estates with semi-detached houses, or small rows of terraces. They are situated next to each other with only a main road separating them (see Figure 7.2). In 1975, the population of Gardenia was about 3,000 and in Stonewall about 2,500; but both have in the intervening years lost about 500 of their respective resident populations.

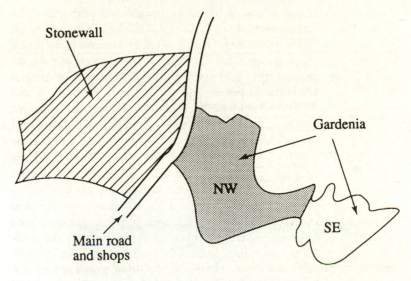

Figure 7.2 Outline map of Gardenia and Stonewall

THE CONTEXT: HOUSING IN SHEFFIELD

Before considering Gardenia and Stonewall in more detail, it is important to explain the general nature of the housing context and the council sector allocation processes within Sheffield. Sheffield is of course a predominantly working-class city, which has been dominated in terms of local politics by the Labour Party since the late 1920s. This has resulted in a particularly strong development of the local authority or council sector within the housing system in the city: in 1981 Sheffield had a much higher proportion of local authority dwellings than the national average (45 per cent as against 32 per cent). During the 1980s Sheffield has continued its somewhat different housing market trends, since it has had only half the national rate of council house sales.

The council house allocation system used in Sheffield is a little unusual and requires some explanation. Unlike many local authorities, Sheffield has never operated a 'points' system of allocation, but instead uses a so-called 'date order' system. To understand this system, Figure 7.3 needs to be examined. A person seeking a council dwelling is placed in one of four categories which are in what John Rawls (1972), in another context, called a 'lexical order', in that, as with a dictionary, the allocator has to exhaust all the '*As*' before

125

A	Urgent clearance Urgent medical Racial harassment, etc.
B	Clearance '*B*' Medical '*B*' Other special priority Modernization schemes
C	Priority transfers (various kinds)
D	General waiting list General transfer list

Figure 7.3   Sheffield council house allocation system 1987

moving to the '*B*s', and so on. However, in considering Figure 7.3 it has to be understood that within the Sheffield council allocation system the *A*s, *B*s and *C*s are relatively small categories, and by far the largest number of people on the housing waiting list at any one time are in Category *D* [1] – which also explains how (given the fact that the system used is a lexical one) Category *D* people can, and in the past did, get allocated to dwellings in substantial numbers. The lexical order, however, operates only for the estate that a prospective tenant puts down as his/her chosen area: hence, there may be fifty or more estates within the city, but when a dwelling falls vacant on a particular estate the allocator does not scan the list of all those in the council housing queue for the city as a whole. Rather, what happens is that the housing allocator goes to the list of people who have put their names down for that particular estate, sees whether there are any *A*s, if not, whether there are any *B*s, and so on. Within each of the *A/B/C/D* categories, the order used is strictly a date-order system. The effect of this system has been that waiting lists for a home in Category *D* (historically the most important category in the system) have varied very greatly as between different estates. Indeed, the Sheffield Housing Department has officially categorized estates into four main categories in terms of their length of Category *D* waiting lists:

1 *low-demand estates* – less than three years
2 *medium-demand estates* – three to five years

3 *high-demand estates* – five to ten years
4 *very-high-demand estates* – ten years plus.

Given this situation, naturally enough people in Category *D* in particular housing need have often been informally steered by housing officers into putting down a lower demand estate as their chosen area, in order to try to ensure a move could be made more quickly.

One final comment should be made on this system. As shown in Figure 7.3, within Category *D* there are two subgroups, namely the 'general waiting list' and the 'general transfer list' (for people seeking to transfer from one council dwelling to another). Since it has usually been the case that people on the transfer list could afford to wait longer for a change of tenancy than could those on the general waiting list, it follows that the proportion of 'transfers' to 'general waiting list' among new residents of a given estate has usually been quite a good indicator of the estate's perceived desirability: the higher the proportion of transfers, the more desirable the estate.

GARDENIA AND STONEWALL, 1975

It is now time to look in more detail at Gardenia and Stonewall as they were in the mid-1970s. Since a full account of these two estates at that time has been published elsewhere (Bottoms *et al.* 1989), only the most significant points will be highlighted here.

Table 7.1 shows that these two adjacent areas had in 1975 quite remarkable demographic similarity. They were not significantly different on any of the variables listed in Table 7.1: social class, sex composition, age composition, percentage married, percentage unemployed, length of stay in current dwelling, and so on. The main characteristics of both areas were that they were very predominantly working class, white, with a high proportion of early school leavers and a high proportion of long-term residents (about 60 per cent of the adult residents of both estates had lived in their current dwelling for more than ten years). Despite these very considerable demographic similarities, however, the two areas had dramatically different official offender and offence rates: Table 7.2 shows that the police-recorded indictable offender and offence rates for Gardenia for 1971 were three times greater than for Stonewall.[2] A victim survey was conducted in the two areas in 1975, and this confirmed a clear offence rate difference between Gardenia and Stonewall. The

Table 7.1 Social characteristics of residents in Gardenia and Stonewall, 1975 and 1988

	Gardenia 1975	Stonewall 1975	Gardenia 1988	Stonewall 1988
Social class of head of household (Registrar General's classification) (%)				
I, II	3	9	3	3
III	57	53	49	45
IV, V	39	37	48	51
Total	100	100	100	100
Sex (all residents in households covered by survey) (%)				
Male	52	49	50	48
Female	48	51	50	52
Total	100	100	100	100
Age (all residents in households covered by survey) (%)				
Under 10	16	10	13	8
10–16	14	12	12	9
17–24	14	14	14	18
25–54	33	35	36	32
55 and older	22	28	25	32
Total	100	100	100	100
% married or living with partner (survey respondents)	67	64	71	59
% unemployed (male survey respondents below age of retirement)	10	11	50	42
% born in Great Britain	97	97	98	98
% white	—	—	97	98
Length of stay in current dwelling (survey respondents) (%)				
Less than 2 years	19	8	15	14
2–5 years	7	15	14	14
5–10 years	14	18	20	12
10–15 years	11	9	15	15
More than 15 years	49	50	36	45
Total	100	100	100	100
Age full-time education completed (survey respondents) (%)				
14, 15, 16	95	96	95	94
Higher age	5	4	5	6
Total	100	100	100	100
Mean household size				
All residents	3.3	3.1	3.0	2.8
Residents aged under 17	1.0	0.7	0.8	0.5

Table 7.2 Police-recorded offender and offence data, 1971 and 1987

	Gardenia 1971	Stonewall 1971	Gardenia 1987	Stonewall 1987
Indictable offender rates per 1,000 population over 10:				
Males	30	7	42	21
Females	5	4	8	6
Total	36	12	50	27
Mean no. of previous convictions of identified offenders	—	—	4.5	4.1
Police-recorded indictable offence rate per 1,000 population over 10 (crimes against residents within area)	31	9	70	41

Table 7.3 Survey offence data (offences within area only)

	Gardenia 1975	Stonewall 1975	Gardenia 1987	Stonewall 1987
A Incidence of offences				
Household offences (mean no. of offences per household)	0.58	0.29	0.59	0.64
Motor vehicle offences (mean no. of offences per m/v owner)	(a)	(a)	0.44	0.47
Personal offences (mean no. of offences per person)	0.05	0.03	0.07	0.10
Overall survey rate per 100 households (uncorrected)	71	37	76	86
Overall *corrected* survey rate (b)	72	37	79	90
% of offences 'serious' (c)	—	—	29	30
B Reporting to the police				
% of offences stated by respondents to have been reported to police (d)	27	24	27	41
Ratio survey data: police data (e)	6.0	6.3	4.2	7.8

Notes: a In 1975 motor vehicle offences were not separately calculated and are included in household offences.

b Corrected to constant 2.7 persons per household aged 10 or over.

c The 'serious/non-serious' distinction was rather crude, basically placing very minor thefts, damage or nuisance type offences into the non-serious category. It was, however, consistently applied between areas.

d In 1975 excludes vandalism and milk theft.

e In 1975 excludes milk theft; also some caution in interpretation is necessary as police and survey data refer to different years.

survey offence rates were different by a ratio of two to one (see Table 7.3), and in fact were different by a ratio of three to one if vandalism was excluded (an important point at that date because low value vandalism was then excluded from the official criminal statistics). The ratio of the survey offence rate to the police-recorded offence rate in the 1970s was also similar for the two estates at six to one, as can be seen from the bottom row in Table 7.3.

Additionally, a self-report study was carried out in 1975 among juveniles attending local secondary schools. This also showed a difference between Gardenia and Stonewall, in the expected direction, especially for frequency as opposed to prevalence of offending: so, for example, theft offences were committed twice as frequently by boys on Gardenia who admitted any theft offence (for further details see Bottoms *et al.* 1989: 47–9). These offending differences between the two estates were also confirmed more qualitatively by a participant observation study carried out in the late 1970s by Polii Xanthos (1981): from her observations, there seemed no doubt that offending behaviour was substantially more frequent among the residents of Gardenia than among those of Stonewall.

These various data presented an obvious problem for explanation: namely, how was it that two such demographically similar areas could have such different offender and offence rates?

In the case of Stonewall there seemed to be little problem in producing a satisfactory explanation for its non-criminal behaviour and reputation. Stonewall had been created as a select estate, for the artisan sector of the working class, in the 1920s, and always thereafter managed to retain a good reputation as a 'respectable' area. It was historically much sought after by families queuing within the Sheffield council house allocation system and, although by the 1970s it had lost some of its earlier attractions because of its age and lack of modernization, it was nevertheless widely regarded as a pleasant estate with good local amenities. It was also officially classified by the city's housing department as a 'high demand' estate with a long waiting list. Most of the families who came to live in Stonewall could be characterized as 'respectable' or 'ordinary working-class' families, with a greater tendency to privatized life-styles, parental supervision of children's leisure time, and effective parental socialization patterns, than was the case in Gardenia (Bottoms *et al.* 1989, especially pp. 52–3). The interaction of the housing market and family and peer social processes thus helped to maintain Stonewall's position as a low-crime estate.

Gardenia in the 1970s was a very different kind of area. Its high offender rate was well known in the city, reinforced by the fact that its street names were of a particular kind which made them easily identifiable when offenders from Gardenia appeared in the local magistrates' court. It was widely perceived by working-class residents of Sheffield as one of the least desirable low-rise housing (as opposed to high-rise flat) developments in the city, and was characterized by the city council as a 'low-demand estate'. Although, like Stonewall, it had begun life early in the twentieth century as a select estate, this was emphatically no longer the case, as one of the older residents complained:

> It used to be select did Gardenia. . . . It was a very select upper working class type of family who got a house up here. Now they're putting in . . . problem families, they've turned the estate into a tip. The language, the state of their houses . . . it's everything. You couldn't leave your door open now when you go out, you could before you know, but now there'd be nothing left.

The adverse reputation of Gardenia in the 1970s had an impact not only on outsiders, but also on some of the residents themselves. There was on occasions a clear sense of shame about admitting to living on such a notorious estate, despite the fact that within the estate itself most, though by no means all, residents felt a sense of belonging (Bottoms *et al.* 1989: Table 5).

Explaining how Gardenia, from a similar starting-point, became so very socially different from Stonewall requires understanding of a series of different factors. Five are especially worth noting (for a fuller account, see Bottoms *et al.* 1989). First, it was clear that Gardenia had 'tipped' in the 1940s, the evidence for this being both anecdotal and statistical. Residents in the area spoke of the area having gone quite rapidly down hill in that period, and we were able to show statistically (from housing department records) that the rate of moving from Gardenia was twice that from Stonewall in the decade of the 1940s (Bottoms *et al.* 1989: Table 4, p. 51). Quite why Gardenia tipped at that date when Stonewall did not is not entirely clear. Locals tended to attribute it to the allocation of a couple of problem families, the argument being that from the allocation of those families a process of downward spiral had begun. It was also the case, however, that at that time there were some newly built estates nearby with better amenities than Gardenia, and it is likely that these estates reduced Gardenia's attractiveness to potential residents. (In this regard it is also worth noting that the physical amenities of the houses in Gardenia at that date were rather less good than those of Stonewall.) The precise history is not now retrievable, but the evidence does suggest a spiralling series of individual decisions to quit Gardenia in the 1940s. It would also seem, although this must be somewhat speculative, that what occurred was quite similar to what Richard Taub *et al.* (1984) have described in their Chicago study, namely that individuals when faced with decisions about whether to move from or stay in an area tend to be quite heavily influenced by what others do.

Second, once the area had tipped the Sheffield housing allocation system operated so as to maintain Gardenia's poor reputation. Incoming tenants in Gardenia in the mid-1970s were mostly those who were either desperate for housing, or those with family links in the area: others in the local authority allocation system avoided the estate if they could. As regards the first group (those desperate for housing), it should be reiterated that Gardenia was widely perceived as one of the least desirable housing (as opposed to high-rise flats) areas in the city, and thus tended to be shunned by all those without special links with the area, unless they felt that, in their particular

Table 7.4 Reasons for accepting present dwelling for tenants in Gardenia and Stonewall in 1975 (at time of original entry) (%)

	SE Gardenia	NW Gardenia	Stonewall
Not really wanted/first possibility of housing/last choice available under housing dept rules/estate suggested by housing visitor	44	29	7
Area of origin/dwelling close to family	22	15	25
Already living on estate	24	26	12
Like area/already living nearby	2	13	22
Slum clearance applicant, near to previous housing	6	14	28
'Convenience' factors (near to town, work or school) or 'want suburbs'	2	1	6
Compulsory transfer	0	2	0
	100	100	100
	(*n* = 46)	(*n* = 103)	(*n* = 105)

Note: No information was available in the following proportion of cases from the full samples: SE Gardenia 23%, NW Gardenia 36%, Stonewall 45%
Source: Sheffield City Housing Department data

housing situation, they could afford only a short wait. As regards the second group (those with family ties in the area) it is worth noting that paradoxically, under the Sheffield council allocation system, relatives could more easily move to a house near their relatives in low demand estates than in high-demand estates. Table 7.4 provides some empirical support for these remarks. As may be seen, many more of Gardenia's tenants in the mid-1970s had reached the estate on a 'Hobson's choice' basis (row 1 of the table) than was the case in Stonewall; slightly more had already lived on the estate (often with their own parents) before their last move, or else wanted to live on the estate because it was their area of origin (rows 2 and 3 of Table 7.4). Further empirical support for the relative popularity of Gardenia and Stonewall in the 1970s is found in Table 7.5: among 'Category D' housing entrants to the two estates (i.e. types (v) and (vi), 'general waiting list' and 'transfers'), Stonewall had a much higher weighting in favour of transfers, and this (as previously indicated) is a good index of desirability.

Table 7.5 Housing allocation by application type 1974 and 1988 (%)

	Gardenia 1974	Stonewall 1974	Gardenia 1988	Stonewall 1988
(i) Clearance	35	75	1	1
(ii) Priority from waiting list (medical/ special hardship, etc.)	17	6	4	8
(iii) Priority transfers				
– intra-estate	} 3	} 6	3	11
– from different estate			8	19
(iv) Homeless list	—	—	41	42
(v) General waiting list	25	3	28	8
(vi) Transfers				
– intra-estate	9	4	6	3
– others	11	6	7	7
(vii) No information	0	0	2	1
	100	100	100	100

Source: Housing Department data

Third, participant observation in Gardenia and Stonewall in the 1970s revealed that Gardenia had a criminal subculture of a mild but thriving kind, especially in the south-east portion of the estate which, because of surrounding geographical features, was physically largely cut off from other local areas, and thus had many of the features of a large cul-de-sac (see Figure 7.2). This criminal sub-culture was inextricably linked with a complex series of strong and interlocking extended family networks (for further details, see Bottoms *et al.* 1989, including the Addendum; also Xanthos 1981).

Fourth, in looking at the explanation for the difference in criminality between Gardenia and Stonewall we have to consider the socialization processes in the two estates, especially with regard to children and adolescents. Participant observation in the two areas in the mid-1970s suggested that in terms of family socialization there was weaker parental supervision and more erratic discipline among the families on Gardenia (Bottoms *et al.* 1989: 52–4, 58; see also Herbert 1976 on family discipline in contrasting council estates in Cardiff); there also appeared to our participant observer to be a very substantial difference between the peer socialization processes in the public spaces of the estates, with a particular tendency for children and adolescents to play outside their houses a great deal more in Gardenia than in Stonewall, and for criminal norms to be transmitted within the context of that play. (See generally Martens 1990, following Bronfenbrenner 1977, on the interaction between parental socialization and the local social environment within which this takes place.) A further possible socialization difference between the two estates was perhaps to be found in a school effect: although we made no direct study of the schools of the area, local reputation had it that the secondary school serving Gardenia at that date had significantly more problems, and a worse school climate, than the equivalent school serving Stonewall (on school effects see generally Graham 1988; Lindstrom 1990).

Finally, the spatial context of the south-east of Gardenia appeared to assist with the maintenance of its criminality. The fact that the area was cut off from other parts of the city by geographical features, coupled with its very adverse reputation, seemed to help it to maintain its rather distinctive and inward-looking social charac-teristics. The official offender rates were higher in this south-east corner than in the rest of the estate, but offence rates were more equal (Bottoms *et al.* 1989: note 20).

Earlier in this chapter we suggested that our central theoretical claim concerning the housing market has to be understood in conjunction with a range of other relevant social and cultural factors, including networks, socialization, social space and reputa-tion. Reviewing briefly the experience of Gardenia and Stonewall in the 1970s this claim can, we think, be supported. The housing market and the allocative processes leading people into Gardenia and Stonewall was the key to the explanation of the differences between the two estates, but this was reinforced and elaborated by these additional social factors.

Finally, however, we must note that Stonewall residents in the 1970s were worried that their estate might be in the process of beginning to 'tip'. A particular anxiety for them was the high percentage of slum clearance tenants moving into the area at that date, following the demolition of quite large privately rented areas of Sheffield in the fairly near vicinity (these were mostly the old 'back-to-back' houses from the nineteenth century, built close to industrial plant). Within the Sheffield council allocation system, slum clearance applicants held a high priority position ('A' or 'B': see Figure 7.3), so relatively 'desirable' estates tended to acquire a higher proportion of such applicants among their new tenants than did low-demand estates at times when slum areas were being cleared. (This outcome was especially likely if the desirable estate was in the same part of the city as the demolition scheme, as Stonewall was in the 1970s.) However, 'slum clearance' families were widely regarded throughout working-class Sheffield as a stigmatic group, despite their favoured position in the official allocation system. In Stonewall, the 1970s influx of such tenants (documented in Table 7.5) therefore led to immediate fears, at least among a minority of residents, of deterioration and 'tipping'. As one resident remarked at the time:

> One neighbour moved to [X estate] and the other got an exchange to [Y estate]. We don't really want to move, this has always been a select estate. You were one up if you lived here. It worries you though wondering who you'll get in as neighbours. I think in the end we'll put in for [X estate] too.

The problem for tenants of local authority estates in Britain, at least in the 1970s, was that although they were players in a housing market they had very limited capacity actually to play that market. Their ability to use the market to leave a given residential area was limited by the difficulty of transfers within the local authority system (see the lowly position of transfer applicants within the Sheffield allocation system: Figure 7.3) and the constraints on moving into owner-occupied tenure created by income, mortgage-worthiness and powerful cultural constraints on changing tenure type. The result of this lack of housing market mobility was that tenants were highly sensitive to any signs that local authority decision-making would lead to a decline in their area, from which they feared they would not be able easily to escape. The allocation to an area of any group who were perceived as signalling such changes, therefore, was greeted with alarm. In the 1970s 'slum clearance' allocations were stereotyped in this way; as we shall see, in the 1980s 'homelessness' allocations were similarly regarded.

GARDENIA AND STONEWALL: THE 1988 REPLICATION

Physically, Gardenia and Stonewall had changed a little by the late 1980s. There had been an extensive programme of modernization carried out by the city council in both estates. This had entailed the demolition of some houses because they were thought to be beyond repair, so one-sixth of Gardenia houses and one-tenth of Stonewall houses had gone completely, and many other people had to be moved out of their houses temporarily while renovation took place. The demolition of houses took rather different geographical forms in Stonewall and Gardenia: in Stonewall it tended to take place rather randomly around the estate, whereas in Gardenia the section most affected by demolition was the middle part of the estate, between the south-east and the north-west sections, leaving these two sections more physically differentiated than before (see Figure 7.2).

Demographically, however, the two estates did not change very much from 1975 to 1988. They were both still very much working class and white, with a high proportion of early school leavers and a high proportion of long-term residents (Table 7.1). However, there were two important demographic changes which affected both estates; both these changes reflected national trends. They were, first, a significant reduction in average household size in both estates, caused particularly by a reduction in the average number of children. (This of course also resulted in a reduced proportion of children among the total population on both estates.) Second, there was a big increase in unemployment in both estates, from about 10 per cent of adult men below the age of retirement in 1975 to over 40 per cent in 1988 (Table 7.1).

In terms of crime and offenders, the official police data still showed differences between the estates (see Table 7.2). According to these data, the offender rate was now twice as high on Gardenia as on Stonewall, and the offence rate had a ratio of 1.7 to 1 across the two estates. These differences were, therefore, reduced from those of the 1970s, but still there were clear differences. However, when the victim survey data for 1987 were examined a rather different picture emerged. The total survey offence rate (see Table 7.3) was at a similar level on both estates, with the rate being, if anything, a little higher on Stonewall.

How could one explain the fact that the police data showed a difference between the offence rates of the two estates, whereas the survey data suggested no difference between them? Could it be, for example, that the public were reporting a higher proportion of offences in Gardenia, thus leading to a higher official offence rate in that area? The answer to this was in the negative, in fact according to victim survey data the reporting rate was rather higher in Stonewall than in Gardenia (see Table 7.3). Was there then a difference in the seriousness of crime, with Gardenia having the more serious crime profile? We examined this possibility by dividing all survey-reported crimes into 'serious' and 'non-serious' crime; again, there was no difference between the two estates (see Table 7.3). The residual possibility therefore suggested was that there was a police recording difference; the possibility of such a difference was heightened when it was observed that the ratio of survey to police data was (by contrast with the 1970s) now noticeably higher on Stonewall than Gardenia (see the bottom row of Table 7.3). Careful local investigation confirmed that there was indeed a recording difference: the two estates were policed by two separate divisions of the South Yorkshire Police, and this had allowed the community constables in each area to develop different practices as to the way in which multiple offences of a similar type were counted. Interestingly, this variation in practice may not be unconnected to the crime reputations of the two estates: in Gardenia the community constables *expected* to record more offences than in Stonewall.

All the evidence suggested, then, that the offence rates of Gardenia and Stonewall were by 1987 similar, despite the considerable differences between them in the mid-1970s. The official offender rates, however, still showed an area difference, and from the evidence that we could gather there seemed no obvious reason why this should be incorrect.[3] Despite this, all the crime indices (including the official offender rate) showed a smaller difference as between the two estates in 1987 than in 1975. Perceptual data from the residents' surveys in these two years also supported this view.

Table 7.6 Residents' perceptions of area

	Gardenia 1975	Stonewall 1975	Gardenia 1988	Stonewall 1988
% feeling at home in area	80	88	83	87
% would be able to get neighbour to shop if respondent ill	—	—	73	74
How many people from other parts of Sheffield would like to move here?				
% many	17	45	22	42
% none	25	6	17	14
Would tenure-holders like to leave area and live somewhere else?				
% yes	35	17	33	29
% responding that living in the area had				
No good features	8	2	—	—
No bad features	21	48	—	—
Area problems identified as 'very big' or 'quite big' (%)				
Vandalism	—	—	62	48
Dogs	—	—	70	51
Litter/rubbish	—	—	60	56
Rowdy kids	—	—	32	27
People fighting	—	—	13	7
Drunken behaviour	—	—	18	12
% of respondents with children believing mixing with wrong sort of kids in area is a problem	—	—	49	20
Has area improved/got worse in last two years?				
Improved	—	—	37	24
Same	—	—	27	44
Worse	—	—	35	32
			100	100
Have recent move-ins changed area for better or worse?				
Better	—	—	15	4
Same	—	—	52	48
Worse	—	—	33	48
			100	100

So, for example, there was a clear reduction in the gap between the two areas in perceptions of area problems (although unfortunately the questions were not quite the same in the two surveys: see Table

7.6); and when we asked identical questions about the extent to which various hypothetical incidents occurred in the two areas (for example, a drunken man on a Saturday night, petty vandalism, etc.) there was again a reduction in the gap between the two estates (Table 7.7). We also asked in the 1988 survey whether recent move-ins into the area had, in the perception of the residents, changed the area for the better or the worse. What was interesting here was that Stonewall residents had a tendency to see recent move-ins as changing the area for the worse to a greater extent than Gardenia

Table 7.7 Residents' perceptions of local offending and policing

	Gardenia 1975	Stonewall 1975	Gardenia 1988	Stonewall 1988
Level of offender rate				
Higher than elsewhere	30	3	26	10
Same	50	55	58	56
Less than elsewhere	6	28	10	13
Don't know	14	14	6	21
	100	100	100	100
Level of offence rate				
Higher than elsewhere	17	3	25	12
Same	33	39	52	38
Less than elsewhere	31	42	13	26
Don't know	19	16	11	26
	100	100	100	100
Chance of being burgled				
Higher than average (%)[a]	25	3	45	37
Hypothetical incidents				
% saying kind of incident occurs 'quite often' (or more)[b]				
Drunk man	46	21	30	23
Theft from supermarket	37	21	27	28
Petty vandalism	86	39	61	52
Husband assaulting wife	24	6	23	10
TV licence evasion	62	24	72	55
Adequacy of police patrolling of local area[b]				
Not enough	67	58	77	65
About right	31	42	23	34
Too much	3	—	—	1
	100	100	100	100

Notes: [a] 1975 and 1988 questions were in different format and cannot be directly compared.

[b] Excludes 'don't know' answers.

residents; while conversely a higher proportion in Gardenia than Stonewall (though still only 15 per cent) were willing to say that recent move-ins had changed the area for the better (Table 7.6).

More research still has to be done on offending in Stonewall in the 1980s, and in particular on whether the offences committed in the area were committed by locals or outsiders. Nevertheless, all the evidence (criminal and social) suggests a significant change in the relative social balance between the two estates in 1987–8 as compared with 1975, so we must now look for some explanation of this change.

Once again, we turned first to the housing market, and it was immediately apparent that Stonewall had not been protected by the local council in terms of housing market pressures from 1975 to 1988. First (as we have already noted), in the 1970s Stonewall received a high proportion of slum clearance tenants, and this seems to have reduced the desirability of the estate in the perception of potential residents in the city as a whole, particularly at a time (in the late 1970s and early 1980s) when other new estates were being built and opened in the city. Second, the Housing (Homeless Persons) Act 1977, with its requirement for local authorities to house homeless families, created a radically new housing market context. The local authority in Sheffield chose to meet its statutory obligations under this Act by treating homeless people as in effect (although not formally) at position D1 in the lexical allocation system (see Figure 7.3; also note 1), except that unlike all other applicants they were made only one offer of housing rather than up to three. Initially, the housing department restricted homelessness applicants within the housing allocation system to 'low-demand' and 'medium-demand' estates (as previously defined), in part because it was felt that allocating them to high-demand estates would constitute unfair 'queue jumping'. More recently this policy was changed, partly because not enough homes were being made available to satisfy the 'homelessness' demand, and partly because it was felt that a principle of egalitarianism meant that all estates in the city should take their share of homelessness allocations. Although this is the formal policy, we know from a separate study of one very-high-demand estate that in fact that estate has continued to be totally protected from homelessness cases. Stonewall, although historically a high-demand estate, did not receive such special protection and as can be seen from Table 7.5 it, like Gardenia, received about 40 per cent of its new tenants in 1988 from the homelessness list. This 40 per cent included some gypsies being moved on to the estate, especially in the winter months. The usual alarm about an estate potentially tipping because of the number of 'undesirables' being received (in this instance homelessness cases) was thus further fuelled by a strong traditional prejudice against gypsies. To the older residents all this added up to the fact that their estate had clearly gone downhill.

We can then postulate that the principal reasons for the decline of Stonewall *vis-à-vis* Gardenia from 1975 to 1988 relate to the changed position of Stonewall in the Sheffield council housing market, and the consequential changes in population composition and in perceptions of the estate by Stonewall tenants and prospective tenants.

Turning now to Gardenia, there is clearly a general feeling (both among residents and among informed outsiders) that Gardenia has shown marginal social improvements during the 1980s. A series of factors seem to be related to this. First, the relative housing market position of Gardenia has improved in view of the rapidly declining popularity of all high-rise flats in the city. Gardenia was always one of the lowest demand housing estates in the city, but with the accelerated unpopularity of large high-rise developments there has been a changed relative perception of the desirability of houses and flats, and Gardenia's absolute position in the housing market has marginally improved. Indeed, in the Sheffield housing department's official categorization of demand levels, Gardenia in 1988 was characterized as 'medium-demand', in contrast to its earlier position as a 'low-demand' estate. Like Stonewall, Gardenia received a relatively high proportion of homelessness cases; but it also took a high proportion of applicants from the general waiting list (indeed a higher proportion than Stonewall: see Table 7.5) and successful general waiting list applicants are now less desperate for housing than they were in the mid-1970s, because people desperate for housing cannot now obtain housing easily through the general waiting list. (This is the result of the interjection of the homelessness category at a higher point in the allocation system.) Thus paradoxically, although Gardenia had roughly the same proportion of new tenants from the general waiting list in 1974 and 1988 (Table 7.5), this group is now a significantly less 'desperate for housing' group than it was in the 1970s. Second, there is some evidence of the break-up of the old criminal subculture in south-east Gardenia, a break-up linked to the modernization and decanting processes of the 1980s. This has also led to some diminution of the strong familial social networks existing in south-east Gardenia in the 1970s, though the evidence of the 1980s is still that Gardenia has a higher offender rate than Stonewall, and stronger family and friendship ties. Third, the school has apparently improved (although again we have not been able to assess this directly). Fourth, in the late 1970s and early 1980s there was a special crime reduction project introduced in Gardenia by a voluntary organization, aimed particularly at young children below the age of 10, and intended to prevent them from taking up a criminal life-style through an introduction to other more desirable social activities. The project itself seems not to have been particularly successful, according to the evaluation of it carried out at the time. However, the project has apparently had a longer-term effect in developing the social skills of a number of residents in the community who have been able subsequently to act in some ways as informal community leaders, and to develop various kinds of social activities on the estate. (As it happens Gardenia has always had more physical resources, such as a school, a pub and a club, which could be used for community development than has Stonewall.) The evidence seems to suggest that this greater degree of resident self-organization has had some impact in improving the self-confidence of residents, and the perceived social desirability of the estate. Fifth, we have noted earlier that there was a reduction in child density in Gardenia (as indeed there was in Stonewall). Reductions in child density would normally be expected to lead to a reduction in offender rates because of the association of adolescents with the peak age of offending. In fact this reduction in child density has probably helped Gardenia much more than Stonewall, since in Gardenia groups of adolescents roaming around the estate were much more of a problem in the 1970s than they were in Stonewall, where the adolescents tended to lead a much more family-centred and private life-style.

At earlier points in this chapter we have referred to the complex interaction of the housing market with other social factors such as cultural networks, socialization, social space and reputations in explaining residential crime careers. This discussion of Gardenia and Stonewall in the 1980s once again seems to confirm that while the housing market is crucial, it has to be understood in close conjunction with these other social processes.

CONCLUSION

In this chapter we have sought to show how the residential community crime careers of two small areas can be analysed, and how changes in the housing market crucially affect the development of community crime careers. Much further work remains to be done in respect of the data in these two areas (and indeed the other four areas in the Sheffield study). In particular it is hoped to model more precisely many of the effects discussed here, that is to say, to develop in more detail an understanding of the precise processes occurring within the model shown in Figure 7.1. In the mean-time it is hoped that this chapter will give a flavour of the ongoing work of the Sheffield research, and an indication of the work to follow.

NOTES

1 In the 1980s the formal allocation system remained unchanged (as shown in Figure 7.3) but, following national legislation, a new 'homelessness' category was introduced, and the city housing department sought to fill 40 per cent of vacancies on most council estates with families or individuals officially designated as homeless. In practice, the homeless list of the housing department was consulted before Category D applicants were considered for allocation – so Category D applicants found themselves in a substantially less favourable position than had previously been the case. The effects of this change on Stonewall and Gardenia are discussed later in this chapter.

2 Because police-recorded crime data in Sheffield had to be collected by hand in the 1970s, it was not practicable to collect data for 1975 as well as for 1971 (see Mawby 1979; Bottoms *et al.* 1987).

3 See Mawby 1979 for a full discussion of relevant issues concerning the measurement of official offender rates in different areas. It is interesting that, among survey respondents with children in 1988, more than twice as many on Gardenia as on Stonewall thought that their children 'mixing with the wrong sort of kids' in the area was a problem (see Table 7.6). On the other hand, the official offender data in 1987 did not suggest that offenders from Gardenia were more recidivistic than offenders from Stonewall (see Table 7.2) which seems impressionistically to be a considerable change from 1975.

REFERENCES

Baldwin, J. and Bottoms, A.E. (1976) *The Urban Criminal*, London: Tavistock.

Bottoms, A.E. and Wiles, P. (1986) 'Housing tenure and residential community crime careers in Britain', in A.J. Reiss and M. Tonry (eds) *Communities and Crime*, Chicago: University of Chicago Press.

—— (1988) 'Crime and housing policy: a framework for crime prevention analysis', in T. Hope and M. Shaw (eds) *Communities and Crime Reduction*, London: HMSO.

Bottoms, A.E., Mawby, R.I. and Walker, M.A. (1987) 'A localised crime survey in contrasting areas of a city', *British Journal of Criminology* 27: 125–54.

Bottoms, A.E., Mawby, R.I. and Xanthos, P. (1989) 'A tale of two estates', in D. Downes (ed.) *Crime and the City*, London: Macmillan.

Braithwaite, J. (1989) *Crime, Shame and Reintegration*, Cambridge: Cambridge University Press.

Bronfenbrenner, U. (1977) 'Lewinian space and ecological substance', *Journal of Social Issues* 33 (3): 199–212.

Coleman, A. (1985) *Utopia on Trial*, London: Hilary Shipman.

—— (1989) 'Disposition and situation: two sides of the same crime', in D.J. Evans and D.T. Herbert (eds) *The Geography of Crime*, London: Routledge.

Giddens, A. (1984) *The Constitution of Society*, Cambridge: Polity.

Gill, O. (1977) *Luke Street*, London: Macmillan.

Graham, J. (1988) *Schools, Disruptive Behaviour and Delinquency*, Home Office Research Study 96, London: HMSO.

Gregory, D. and Urry, J. (1985) *Social Relations and Spatial Structures*, London: Macmillan.

Herbert, D.T. (1976) 'The study of delinquency areas: a geographical approach', *Transactions: Institute of British Geographers* 4: 472–92.

Lindström, P. (1990) 'Pupil, school and delinquency', in P.-O.H. Wikström (ed.) *Crime and Measures against Crime in the City*, Stockholm: National Council for Crime Prevention.

Martens, P.L. (1990) 'Family, neighbourhood and socialisation', in P.-O.H. Wikström (ed.) *Crime and Measures against Crime in the City*, Stockholm: National Council for Crime Prevention.

Mawby, R.I. (1979) *Policing the City*, Westmead: Saxon House.

Merrett, S. (1979) *State Housing in Britain*, London: Routledge.

Morris, T.P. (1957) *The Criminal Area*, London: Routledge.

Newman, O. (1972) *Defensible Space*, New York: Macmillan.

Rawls, J. (1972) *A Theory of Justice*, London: Oxford University Press.

Reiss, A.J. (1986) 'Why are communities important in understanding crime?', in A.J. Reiss and M. Tonry (eds) *Communities and Crime*, Chicago: University of Chicago Press.

Rex, J. and Moore, R. (1967) *Race, Community and Conflict*, London: Oxford University Press.

Taub, R.P., Taylor, D.G. and Dunham, J.D. (1984) *Paths of Neighborhood Change*, Chicago: University of Chicago Press.

Taylor, R.B. and Gottfredson, S. (1986) 'Environmental design, crime and prevention: an examination of community dynamics', in A.J. Reiss and M. Tonry (eds) *Communities and Crime*, Chicago: University of Chicago Press.

Xanthos, P. (1981) 'Crime, the housing market and reputation', unpublished PhD thesis, University of Sheffield.

8

CRIME AWARENESS AND URBAN NEIGHBOURHOODS

David T. Herbert and Judy Darwood

With the advent of national crime surveys in both the United States and the United Kingdom, much improved databases have allowed themes such as awareness of crime and fear of crime to be recognized more fully as research fields. Governmental concern is demonstrated by the Home Office working party on fear of crime (Grade Report 1989) and by the earlier Figgie Reports (1980) which arose out of public and corporate concerns in the United States. The facts which continue to emerge are often startling. The British Crime Survey (Maxfield 1984) found that 12 per cent of all inner city dwellers never went out at night because of fear of crime; the Islington Crime Survey (Jones *et al.* 1986) reported that more than half the women residents in that area avoided going out at night because of fear of attack. This last finding points to the fact that fear of crime has differential impacts and it is women, elderly people and the single households who report most fear. Evidence from the American National Crime Survey on fear of crime is equally persuasive and Hindelang *et al.* (1978) stated that 46 per cent of those interviewed in eight American cities had made adjustments to their normal activities because of fear of crime. More qualitatively:

> Many people, even in the modern and affluent Western world, are haunted by fear. Almost daily we read about muggings and murders, and about elderly residents of the inner city so afraid that they are virtually prisoners within their own home.
>
> (Tuan 1979: 209)

Statements of this kind refer to some of the extreme situations in which fear of crime arises – the most vulnerable elements of the population in the highest risk areas of the city. Much of the literature is concerned with these groups and areas *in extremis* and the Islington Crime Survey is one such example. Fear of crime and awareness of crime, however, have much wider effects and reach out to many different types of people in a variety of living conditions. The empirical component of this chapter is concerned with areas where crime rates are not particularly high but afford the normal range in a middle-sized British urban area. The main thrust of this study will be to assess awareness of residential crime in a set of eight study areas and to examine the effect of this awareness upon selected aspects of people's lives. A general supposition is that a heightened awareness of crime or a perceived high level of crime leads to more worry and greater fear; that these conditions of awareness, worry and fear are closely related and interdependent. Most of the literature is concerned explicitly with fear of crime with implicit assumptions about awareness.

> Surveys enquiring about the extent of disorder problems find responses are closely related to fear of crime and perceptions that serious crimes are neighbourhood problems.
>
> (Skogan 1988: 49)

There are 'signs of crime' in urban neighbourhoods which include both physical decay and unsocial behaviour, in Skogan's (1988) view these conditions generate fear because they signal that the community is out of control.

RECENT RESEARCH ON AWARENESS AND FEAR OF CRIME

Information on the extent of fear of crime begs many questions. Maxfield (1984) posed the question 'Who is afraid?' and concluded that it was misleading to regard fear of crime as a national problem. Some groups feel much more vulnerable than others and the fear of crime which they hold may not bear much relationship to their likelihood of actually falling victim. Fear itself is not an unambiguous feeling and may be compounded from fears of personal safety, loss of property and concerns about the safety of family and friends. One distinction of significance to the empirical study that follows is that made by Maxfield (1984) when he argued that fear relates to street crime whereas worry arises from concern about household crime and burglary in particular. Again fear is by no means based upon direct experience of victimization, it is more often an anticipated fear which can be fed from a number of sources.

Attempts to form models of fear of crime have had limited value. Baumer (1985) included three components of individual characteristics, perception of threat in the locality, and knowledge of local criminal events but these did not prove consistently useful predictors of fear of crime. Taylor and Hale (1986) tested the effects of indirect victimization, disorder and community concern but this model lacked generality.

Most research has tried to identify the kinds of people who are more likely to experience fear of crime and here there is clear evidence for the importance of social and demographic differences. Demographic groups most in fear – elderly people, women and single households – have already been noted but Maxfield (1984) argued that both victimization and fear were highest among the poor. This statement cannot be taken as a necessary truth, the circumstances in which people live have a profound effect and wealthier people may have more worries about some types of crime, such as burglary, than poorer people. It is clearer that the poor suffer more crime as part of the relative disadvantages under which they live; it is also evident that the group most at risk from personal attack comprises young adult males, often from poorer backgrounds.

The relationship of victimization to awareness of crime is generally not regarded as being clear. Smith (1986) suggests that fear of crime can be shown to be more closely related to victimization once allowance is made for people's exposure to risk and their physical vulnerability; this can be summarized as a 'life-style-exposure to risk' thesis. Young adult men are more often in situations where violence can occur but regard this as part of their life-style and are less fearful; elderly people are much more fearful but because they closet themselves away they are far less often at risk. Smith (1989) argues that variations in activity patterns rather than demography may account for differences in rates of victimization among groups of people. Further:

> In sum, victim studies suggest not only that crime is spatially coincident with the most common indicators of urban deprivation, but also that socially, the incidence of victimisation is ultimately connected with the routine practices of an urban life-style.
>
> (Smith 1989: 277)

In the case study which follows the relationship between experience of victimization and perception of crime will be explored, but the evidence of previous research literature is that the connection is likely to be ambiguous and confounded by other sources of variation. One exception to this is the study of Skogan (1987), who found that recent victimization was consistently related to measures of worry and concern about crime. Such recent victims believe that crime is more prevalent in their neighbourhoods and take measures to protect themselves. One of the more productive lines of research into crime and community has been that which examines the effect of neighbourhood disorder and the presence of 'incivilities' upon awareness and fear of crime. Wilson and Kelling (1982) examined the relationship between crime and incivilities or the signs of neighbourhood deterioration. Incivilities can take the form of visible signs in the urban environment such as broken windows, litter, scars of vandalism and graffiti or behaviour such as drunkenness, youths loitering to no perceived good purpose and verbal harassment of passers-by. For Wilson and Kelling, as incivilities increase so a sense of community begins to break down; this in turn may make the neighbourhood more vulnerable to crime. With this comes greater awareness of crime within the neighbourhood and a heightened sense of fear. Hope and Hough (1988) examined the relationship among levels of perceived incivilities, worries about crime, social cohesion within neighbourhood, satisfaction with neighbourhood and rate of victimization. They found the correlations among these to be so strong that 'at the neighbourhood level, the perception of incivilities and self-reported fear of crime and neighbourhood satisfaction might be thought of as equivalent indicators' (Hope and Hough 1988: 36). They were confident that perception of incivilities is a sensitive indicator of a related set of worries about neighbourhood.

One of the Hope and Hough indicators was satisfaction with neighbourhood as a place to live. They found significant negative correlations from their analysis of British Crime Survey data which ranged from −0.73 to −0.90 between perceived incivilities, (−0.90), fear of crime (−0.85) and victimization rate (−0.73) with levels of neighbourhood satisfaction. Satisfaction with neighbourhood as a place to live is diminished if threatened by high crime rates actual or perceived. Using the ACORN classification of British residential areas (see Hope and Hough 1988) it was clear that as rates of burglary, fear of crime and perceived incivilities *increased*, so neighbourhood satisfaction declined. In a study of a group of elderly people, Yin (1982) found that levels of neighbourhood satisfaction and general morale were inversely related to awareness and fear of crime. Several researchers have stressed the ways in which neighbourhoods can deteriorate over time and progressively reduce levels of satisfaction. Skogan (1985) emphasized that this was a reciprocal relationship. On the one hand there are factors which trigger neighbourhood decline and amplify the conditions under which worries of crime grow, on the other these heightened worries can in themselves contribute to declining neighbourhood satisfaction. Rumours of nearby crimes and victims, physical decay and signs of social disorder can all contribute to the start of a downward spiral. Hope and Shaw (1988) reviewed ways in which this spiral might be halted by prompt refurbishment of environment, persuading potential leavers to stay and encourage external investment and intervention.

Awareness of crime may be increased by these 'signs' offered by various forms of incivilities, by direct experience of victimization and by hearing about nearby crime or victims among similar kinds of people. It can also be affected by the media and other agencies, including the police, who may make known the fact that crime is prevalent. These form a variety of social 'cues' some which appear to act independently of risk to convey the threat of crime. Smith (1985) argues that the reporting practices of the provincial press are more important as ways of heightening public awareness about crime rather than of inciting their fears. Others have argued (see Smith 1989) that the mass media in the form of newspapers, television, gossip and rumour are known to exaggerate both the extent of crime and its severity.

As awareness and fear of crime has differential societal impact so it has an uneven distribution in both time and space. For those people who feel most vulnerable, the hours of darkness exaggerate their anxieties. This despite the fact that there is no simple correlation between these times and the actual incidence of crime. Again, awareness and fear of crime is much more concentrated in some kinds of places: it has clear geographical expressions. Hope and Hough (1988) argue that there is strong evidence from the British Crime Surveys for what they term the 'pocketing' of crime or its concentration in particular types of residential area. They use the ACORN classification of residential areas to identify areas of high, medium and low risk and show, for example, that high-risk areas, such as poor public sector estates, have 12 per cent of the households but 37 per cent of the burglaries, 12 per cent of the adult population (over 16 years) but 33 per cent of the robberies. It is the people who live in high-risk areas who are more aware of crime and who feel most vulnerable. 'The poor tend to live in the more dangerous neighbourhoods and are less able to secure their homes and find it more difficult to avoid dangerous areas' (Baumer 1985: 241).

Fear of crime has become a well-studied phenomenon. This is not to argue that it is now well understood. Some have argued that the complexities are such that it may never be. Garofalo (1981), for example, argued that from a purely scientific standpoint, research on the fear of crime can continue indefinitely. This is a sweeping statement but it does point to the need for a strategy to study fear and awareness of crime in its many facets. One lacuna is a fuller understanding of the role of geographical space and the need for more research on this topic is clear. Skogan and Maxfield (1981) argued that fear of crime varied more significantly among neighbourhoods within cities than among different cities. Lewis and Maxfield (1980) argued that attitude studies of residents within urban neighbourhoods on the issue of fear of crime were not common. Maxfield (1984) stated that one shortcoming of the British Crime Survey was its inability to draw any conclusions about levels of fear at a neighbourhood level. The Grade Report (1989) refers to area differences within cities and Table 8.1 which is extracted from that report summarizes some of these variations. Both fear of crime and actual risk are lowest in the affluent suburbs and are highest in the poorest council estates. Higher status areas, types 1 and 3, have the lowest levels of fear though one of these (3) is classified as a high-risk area. The report also showed that although there are high risk areas, levels of fear do not mirror risk. 'The consequences of being burgled may well be easier to accept for those living in well-insured, affluent areas than for those close to the breadline' (Grade Report 1989: 14).

In the empirical study which follows, differences among neighbourhoods will be explored in order to fill something of this research need which the literature has identified.

STUDY AREA

This empirical study was based upon West Glamorgan and adjacent parts of Dyfed. Eight areas were used, five of which were residential

Table 8.1 Area types and victimization and fear rates

Type of area[a]	Burglary		% stating crime as worst feature of area
	% victims in past year	% very worried about next year	
1 Affluent suburbs	3	18	3
2 Poor quality old terraced housing	4	29	6
3 High-status non-family[b]	10	25	7
4 Less well-off council estate	4	31	8
5 Poorest council estate[b]	13	41	17
National	4	23	5

Source: Grade Report (1989: 57) based on British Crime Survey.
Notes: [a] The areas are from the ACORN classification, which stands for A Classification of Residential Neighbourhoods. It is based on census small area statistics and there are thirty-eight area types aggregating to eleven neighbourhood groups.
[b] Classified as high-risk areas.

districts within the Swansea urban area, one was a commuter village in Gower and two were small, former industrial towns (AMM and GORS). The five urban neighbourhoods were selected from a set of planning cells (see Herbert and Hyde 1985); in each of the eight areas a random sampling technique was used to conduct a social survey which produced sample populations ranging from 60 in AMM to 81 in LA2. The urban neighbourhoods comprised two local authority estates (LA1 and LA2) of similar social–demographic characteristics, two inner-city terraced row areas (IC1 and IC2) of which the former was mixed and relatively transient while the latter was uniform, stable and dominantly low income, and a suburban area (SUB) composed of high-income owner-occupiers mainly in detached housing.

The term 'neighbourhood' will be used in relation to each area though it is recognized that it is most appropriate to the urban areas. In each of the study areas questions on local activity systems and cognitive images were used to establish that they could be regarded as 'places' with which people identified. Results from this part of the exercise will not be detailed here but suffice to say that each study area did emerge as a recognized place. From this more widely based study, a number of themes will be extracted which relate to fear of crime. For these some basic hypotheses will be stated and tested against the empirical database which was collected.

151

DATABASES

The three hypotheses which are to be tested all concern the actual level of residential crime within the neighbourhoods, as derived from police statistics and also the level of such crime as perceived by the residents. This involves a comparison between the official reality of the incidence of crime and the imagined level of local people. As police records were used for the initial data collection, this stage of the exercise was vulnerable to the well-known caveats associated with official statistics on crime (see Herbert 1982; Hough and Lewis 1989). Recent surveys such as the Islington Crime Survey (Jones *et al.* 1986) have indicated relatively high levels of reporting and recording of burglary and theft. The Swansea study did also record experience of victimization within the study areas. Some basic points on the data sources and the methodology used will now be summarized.

Details on 812 cases of residential burglary were collected for Swansea areas in 1980, together with 142 cases of burglary from other residential premises and 275 cases of theft. This set of 1,229 cases over the five areas was classed as residential crime. For the three areas outside the city of Swansea, data were collected over a two-year period, 1981–2, in order to provide good-sized samples and this yielded a set of 322 cases. For each case there were details on the offence itself, the victim and the offender where known. One basic procedure was to convert data to normalized crime rates (Clarke 1984) using the formula:

$$\text{ANCR} = \frac{\text{total number of victimized targets}}{\text{total number of potential targets}} \times 10^n$$

expressed per 1,000 households

where ANCR is the average normalized crime rate

A second procedure was to measure the average perceived level of burglary (APBL). This was expressed as:

$$\text{APBL} = \frac{3n_3 * 2n_2 * 1n_1 * 0n_0}{N}$$

where $N = n_3 + n_2 + n_1 + n_0$

or the total number of respondents in the sample. The multipliers are *3 for respondents who believe that burglary is very common in these areas, *2 for fairly common and *1 for not very common.

APBL is a summary statistic of perceived burglary level within each area which produces scores within the range 0.0 to 3.0. For the eight study areas the APBL scores ranged from 0.98 to IC2 to 2.42 in LA1.

HYPOTHESES

The first hypothesis was concerned with awareness of crime in the residential areas and examined the relationship between actual and perceived crime levels:

The perceived levels of neighbourhood crime are positively related to the real levels of neighbourhood crime.

The hypothesis is stated in this more general way in order that it could be applied to different types of crimes but for this section the connection examined is between APBL and ANCR as earlier defined. Table 8.2 shows some basic statistics and Figure 8.1a shows the relationship in a graphical form. There is a close relationship between real and perceived levels of burglary which, as measured by a correlation coefficient of $+0.918$, is significant at the 0.01 per cent level. From this evidence the first hypothesis can be supported and residents have an accurate awareness of the residential crime rates within their own neighbourhood. This is not of course a numerical awareness but one which places local crime at a relative level. APBL is a composite index but if a simpler measure of perceived burglary is used, the percentage regarding it as very common in their neighbourhood, a high correlation ($+0.962$) is still obtained. Within the five urban neighbourhoods, LA1 and LA2 have the highest scores on both actual and perceived levels of burglary; IC1 and IC2 have perceived levels which are below the reality and SUB *vice versa*

Table 8.2 Some basic indicators by residential areas

Residential burglary	AMM	GOWER	GORS	LA1	LA2	SUB	IC1	IC2
Actual resid. burglaries	7	7	16	138	51	24	42	16
Normalized rate ANCR	1.6	2.2	3.6	57.2	34.0	15.5	22.1	10.9
% think burglary very or fairly common	37	22	30	84	68	48	39	10
APBL	1.4	1.2	1.3	2.4	2.0	1.5	1.5	1.0
Ranks on ANCR	8	7	6	1	2	4	3	5
Ranks on APBL	5	7	6	1	2	3	3	8

As Figure 8.1b shows, SUB does in fact have low perceived levels of crime in all categories except burglary showing that it is this offence in particular which concerned high-income suburban dwellers.

Information was available to test the relationship between perceived level of burglary and experience of victimization. The victimization rates revealed by the survey ranged from 15 per cent in GOWER to 66.7 per cent in LA1 which were significantly higher than rates suggested by official statistics though the relative vulnerability of the five city areas, for which accurate comparisons could be made, was contained in identical rank order for ANCR and victimization rates. There was also a high positive correlation (+0.967 per cent) between APBL and experience of victimization at a neighbourhood level. Where respondents stated a victimization

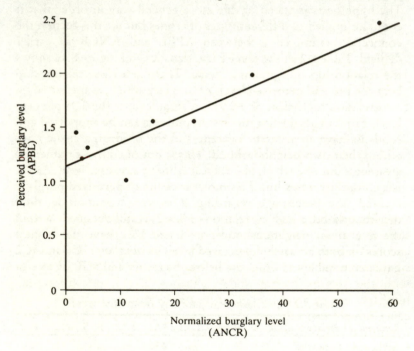

$$y(APBL) = 1.13909 + 0.02173 \ X \ x(ANCR)$$

$$r = 0.918 \text{ significant at } 0.01\% \text{ level}$$

Figure 8.1 Burglary and neighbourhood problems: a comparison of perceptions and realities

(a) Real and perceived levels of burglary

154

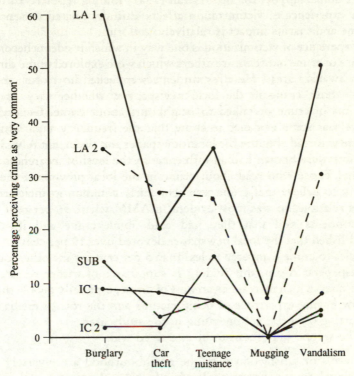

(b) Perceptions of problems in the city neighbourhood

experience of burglary within the previous eighteen months, subsequent analysis examined differences in attitudes between victims and non-victims. For the four urban neighbourhoods which had significant numbers of burglary victims there were large differences with the victims having much higher levels of perception of burglary. In one area, LA1, both victims and non-victims (62 per cent) thought that burglary was very common and this appeared to override the significance of victimization experience in that neighbourhood. Generally there were many more non-victims in the 'don't know' category and this seems to support Hindelang *et al.* (1978) who argued that the main effect of the victimization experience was to enhance *awareness* of crime. For the three non-Swansea areas, a composite residential crime index showed significant differences between victims and non-victims, using the Mann Whitney U-test for AMM and GORS. Overall, then these results offer some support for the Skogan (1987) finding reported earlier that experience of victimization affects attitudes related to fear of crime and that its impact is relatively uniform.

Experience of victimization is one way in which residents become aware of crime but there are others which were explored in the three non-Swansea areas. Here respondents were asked how often they read about crime in the local newspapers, whether they knew victims of crime or talked to neighbours about crime. Generally there was some evidence to show that the frequency with which crime was read about in the local newspaper and the extent to which victims were known had an influence on the level of awareness of crime. Those who read about crime in the local press were more likely to believe that crime was very fairly common in their area. This relationship was most evident in AMM where 56 per cent of respondents said that they had read about crime and it was established that the local newspaper devoted over 10 per cent of its articles to crime compared to less than 5 per cent for the other local newspapers. Again, this finding is supportive of earlier research. The 'cues' which increase awareness of crime, especially people they know having experience of being victims and the role of media in reporting offences, are operating in the study area.

The second hypothesis to be examined was

Levels of satisfaction with the neighbourhood are inversely related to the perceived levels of criminal activity within neighbourhood.

The perceived quality of life within residential neighbourhoods is affected by a variety of criteria of which awareness of crime is one. The Grade Report (1989) shows that crime is rarely the main criterion which residents use as even in the highest risk areas it tends to be placed behind considerations such as lack of jobs and local facilities. A Likert scale on six attitudinal variables was compared with a direct question on neighbourhood satisfaction and as the correlation was high the direct question was used as the basic measure. Figure 8.2 shows the relationship between levels of satisfaction in the neighbourhood as a place to live and perceived levels of local crime. There was a strong negative correlation of -0.84 which is significant at the 0.01 per cent level. If the inverse relationship is measured by relating those dissatisfied with APBL, a correlation of $+0.73$ is obtained. As another test, respondents were classified into those who (involuntarily) mentioned a problem of

(a) Perceived levels of burglary and satisfaction

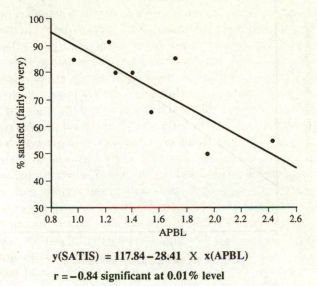

$$y(SATIS) = 117.84 - 28.41 \times x(APBL)$$

r = −0.84 significant at 0.01% level

(b) Perceived levels of burglary and dissatisfaction

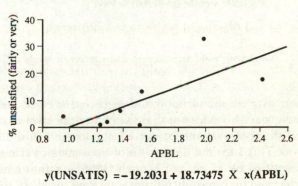

$$y(UNSATIS) = -19.2031 + 18.73475 \times x(APBL)$$

r = +0.730 significant at 0.5% level

Figure 8.2 Neighbourhood satisfaction and perceived levels of burglary

any kind or made some sort of negative assessment of their neighbourhood and those who did not. The correlation between satisfaction and not mentioning any problems was +0.95. In summary, these tests confirm a strong negative relationship between

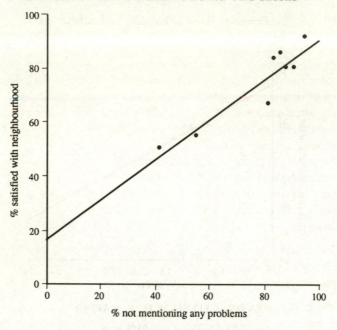

$$y(SATIS) = 15.104 + 0.759 \quad X \quad x \text{ (NO PROBS)}$$

$r = 0.945$ significant at 0.01% level

(c) Absence of problems and satisfaction

Figure 8.2 Neighbourhood satisfaction and perceived levels of burglary (cont.)

satisfaction with neighbourhood and perceived level of residential crime; there was also evidence that the higher the level of criminal activity in the area, real and perceived, the lower was the level of satisfaction. This adds the dimension of awareness of crime to the cluster identified by Hope and Hough (1988) – rates of burglary, fear of crime and perceived incivilities – as being inversely related to neighbourhood satisfaction.

As a final line of inquiry on the theme of satisfaction, differences between victims and non-victims were examined with the expectation that victims would be less satisfied. In all but the two higher-status areas, SUB and GOWER, victims were less satisfied. These two areas returned no dissatisfied residents and the positive qualities of these areas may offset any impact of a victimization experience.

When the six areas (excluding SUB and GOWER) were further tested by separating the two groups of victims and non-victims there were limited statistical differences and it appears that although direct experience of victimization has a negative impact, it is only one of a number of such factors. Earlier discussion of the various forms of incivilities is relevant here and as Sparks *et al.* (1977: 212) argue, 'attitudes towards the neighbourhood influence perceptions of crime there (and feelings that the neighbourhood is unsafe) rather than the other way around'.

The final hypothesis to be examined related to feelings of vulnerability and to ways in which people react by adopting security devices of various forms. This hypothesis was more explicitly concerned with worry and fear rather than awareness. It was stated:

> The worry about being burgled and the number and frequency of the security measures taken are positively related to the perceived risk of being burgled.

For the first part of the analysis the Swansea and non-Swansea study areas have to be separated as the theme was pursued in rather different ways in the two sets of areas. In AMM, GOWER and GORS one question was whether or not respondents ever worried about being burgled: the positive responses were 36.6 per cent, 31.3 per cent and 36.6 per cent respectively. Ensuing analysis based on the expectation that worry would lead to greater perceived risk was not conclusive in terms of its results. This finding is not at odds with the literature, Garofalo (1981) found that fear of crime is not a simple reflection of risk or of experience of victimization and that the more effective measures of fear relate to threat to person rather than to property. Overall the most common effect of worry about being burgled was to increase wariness of strangers. Respondents in GORS were most affected by worry and those in AMM least despite the fact that the latter tended to be elderly. Other evidence showed that whereas AMM was a socially integrated and supportive community, GORS was at the opposite end of the spectrum. In each of these three areas, respondents were asked which types of people they thought most likely to be burgled and the rankings from single elderly people at one end to families with children at the other reflected these perceived levels of vulnerability for different demographic types.

For the five city areas, the information on worry was not

available but a more specific question, 'How safe do you feel walking around this neighbourhood after dark?', was asked. Generally the feeling of lack of safety was high with between 23.8 and 50.0 per cent of respondents feeling unsafe. LA1 had the highest score and IC2 the lowest, a range which does fit with their position in 'vulnerable' or 'risk' space. SUB was the exception with a much higher score than could be expected from real or perceived levels of risk. Questions on the frequency with which youths were seen loitering or strangers were encountered in the neighbourhood threw some light on these scores. Over 80 per cent of those in SUB never or rarely saw youths loitering but 24 per cent often saw strangers. This may fuel the apparent tendency of higher income groups to have higher perceptions of risk than reality justifies, but it is worth noting that this awareness of an 'incivility' did nothing to diminish satisfaction with this affluent suburban area as a place to live.

Data collected to measure security consciousness were concerned with security of dwelling and not of person or neighbourhood; measures related to actions taken when the dwelling was left unoccupied such as locking door, closing windows, informing neighbours, cancelling deliveries and leaving on lights. A security consciousness index was compiled from four indices – locking front door, locking back door, closing downstairs windows and leaving on a light at night, with scores based on the frequency with which these measures were adopted. When mean measures on the index were compared with APBL, see Figure 8.3, there is a positive correlation of +0.793 which is significant at the 0.05 per cent level. There was also evidence that security consciousness is related to respondents perceived risk of people like themselves being burgled.

In the five city areas a considerable amount of information was collected on security hardware in dwellings. SUB was the best protected area followed by IC1; the other four areas were all poorly protected. This pattern reflects ability to afford protection rather than actual crime risk and the finding supports Skogan (1981: 34): 'Household protection seems to be an economic act unrelated to the direct threat of crime'.

Questions were asked on the perceived adequacy of locks to deter children, teenagers, young burglars and professional thieves. All areas showed a regular gradation with SUB, for example, having a score of 94 per cent for children and 24 per cent for professional thieves. SUB had the greatest faith in locks as a deterrent, LA1 the least. Generally the preference was for 'hard' security such as locks

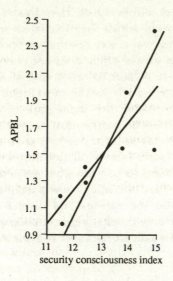

(1) y(APBL) = -1.97595 + 0.26753 X x (SEC)
    r = 0.793 significant at 0.05% level

(2) y(SEC) = 9.52 + 2.35 X x(APBL)

Figure 8.3 Perceived levels of burglary and security consciousness

and alarms rather than for social measures such as reliance on neighbours or police.

CONCLUSIONS

This study has examined three hypotheses related to worry about burglary and to the broader concepts of awareness and fear of crime. The eight study areas have allowed neighbourhood differences to be examined in a systematic and rigorous way. These neighbourhoods are drawn from a part of the United Kingdom which has overall unexceptional levels of crime but serves to show that even within such a locality the area variations can be of some significance. Any generalizations about the incidence of awareness and fear of crime clearly need to be modified by considerations of differences in geographic as well as social space; these along with actual crime rates are 'pocketed' in particular kinds of residential areas. People generally have a good awareness of the relative levels of crime risk of the areas within which they live. Their level of satisfaction with the places in which they live is adversely affected by these perceptions in real ways. Even in the worst of these areas, however, worry about residential crime does not override all other considerations and it has to be seen as one strand in the quality of life assessment and not necessarily as the most significant strand. The socio-economic status and the demographic variables are always significant and those who worry most are not always those with most cause. Attitudes towards the need for security and actual adoption of security measures bear a relationship with perceived levels of crime but the ability to pay for what are often expensive devices is a telling factor. Fear of crime is now part of the everyday lives of many people who for various and diverse reasons feel vulnerable. Where they live, with what kinds of people and in what kinds of residential area has a direct relevance to their awareness and fear of crime.

REFERENCES

Baumer, T.L. (1985) 'Testing a general model of fear of crime: data from a national survey', *Journal of Research into Crime and Delinquency* 22: 239–55.

Clarke, R.V.G. (1984) 'Opportunity-based crime rates', *British Journal of Criminology* 24: 74–83.

Figgie, H.E. (1980) *The Figgie Report on Fear of Crime: America Afraid*, ATO inc., 4 volumes.

Garofalo, J. (1981) 'The fear of crime: causes and consequences', *Journal of Criminal Law and Criminology* 72: 839–57.

Grade, M. (1989) *The Grade Report on the Fear of Crime*, London.

Herbert, D.T. (1982) *The Geography of Urban Crime*, London: Longman.

Herbert, D.T. and Hyde, S.W. (1985) 'Environmental criminology: testing some area hypotheses, *Transactions: Institute of British Geographers* 10: 259–74.

Hindelang, M.J., Gottfredson, M.R. and Garofalo, J. (1978) *Victims of Personal Crime*, Cambridge, Mass.: Ballinger.

Hope, T. and Hough, M. (1988) 'Area, crime and incivility: a profile from the British crime survey', in T. Hope and M. Shaw (eds) *Communities and Crime Reduction*, London: HMSO.

Hope, T. and Shaw, M. (eds) (1988) *Communities and Crime Reduction*, London: HMSO.

Hough, M. and Lewis, H. (1988) 'Counting crime and analysing risks: the British Crime Survey', in D.J. Evans and D.T. Herbert (eds) *The Geography of Crime*, London: Routledge.

Jones, T., Maclean, B. and Young, J. (1986) *The Islington Survey*, Aldershot: Gower.

Lewis, D.A. and Maxfield, M.G. (1980) 'Fear in neighbourhoods: an investigation of the impact of crime', *Journal of Research into Crime and Delinquency* 17: 160–89.

Maxfield, M.G. (1984) *Fear of Crime in England and Wales*, Home Office Research Study 78, London: HMSO.

Skogan, W.G. (1981) *Issues in the Measurement of Victimisation*, Washington, DC: US Department of Justice.

—— (1985) 'Fear of crime and neighborhood change', in A.J. Reiss and M. Tonry (eds) *Community and Crime*, Chicago: University of Chicago Press.

—— (1987) 'The impact of victimisation on fear', *Crime and Delinquency* 33: 135–54.

—— (1988) 'Disorder, crime and community decline', in T. Hope and M. Shaw (eds) *Communities and Crime Reduction*, London: HMSO.

Skogan, W.G. and Maxfield, M.G. (1981) *Coping with Crime*, Beverly Hills, Calif.: Sage.

Smith, S.J. (1985) 'News and dissemination of fear', in J. Burgess and J. Gold (eds) *Geography, the Media and Popular Culture*, London: Croom Helm.

—— (1986) *Crime, Space and Society*, Cambridge: Cambridge University Press.

—— (1989) 'Social relations, neighbourhood structure and fear of crime in Britain', in D.J. Evans and D.T. Herbert (eds) *The Geography of Crime*, London: Routledge.

Sparks, R.F., Genn, H. and Dodd, D.J. (1977) *Surveyng Victims*, Chichester: Wiley.

Taylor, R.B. and Hale, M. (1986) 'Testing alternative models of fear of crime', *Journal of Criminal Law and Criminology* 77: 151–89.

Tuan, Y.F. (1979) *Landscapes of Fear*, Oxford: Blackwell.

Wilson, J.Q. and Kelling, G.L. (1982) 'Broken windows: the police and neighborhood safety', *Atlantic Monthly*, 255: 29–38.

Yin, P. (1980) 'Fear of crime among the elderly', *Social Problems* 27: 492–504.

9

DIFFERENT WORLDS
The spatial, temporal and social dimensions of female victimization

Kate Painter

A FRAMEWORK FOR THE ANALYSIS OF SPACE, TIME AND CRIME

Legal and social definitions of crime do not always coincide and can conflict. Crime is a breach of the criminal law but it also involves a social relationship between victim and offender which takes on particular patterns and meaning in space and time. Time and space provide the setting for all criminal interactions, the setting being essential to specifying the context of the criminal act. In short, space and time are crucial to any understanding of how crime is generated, explained and, above all, prevented.

So far, so trite. At first glance this statement seems to say no more than crime occurs to someone, at some time, in some place. Yet, Giddens (1985) has pointed out that the concepts of space and time have been used by social theorists as 'mere environments' for social action whereas they ought to be regarded as of the utmost importance in the conduct of empirical research in the social sciences. Despite Giddens's observation and the request by Gregory and Urry for a realignment of social theory and human geography which would result in 'a major renaissance of the interconnections between social relations and spatial structures as the central focus for scientific enquiry' (Gregory and Urry 1985: 3), the difficulty remains as to how best to use the concepts of the 'spatial' and the 'social' to analyse crime and crime control in a way which does not imply that one or the other has determinate effects. Though the correlates of space and crime have been recurrent themes throughout the history of criminology (Mayhew 1861; Quetelet 1869; Park *et al.*, 1925; Shaw and McKay 1942; Morris 1957; Baldwin and Bottoms 1976; Mawby 1986; Bottoms and Wiles 1988) it is fair to say that the interaction between the social and space–time dimensions of crime have not been developed in any substantial sense.

The problem is both theoretical and empirical. Much remains to be done to conceptualize the ways in which individuals' lives are constructed within historical processes (including time historical and time present) and the ways in which these connect to their risks and fears of crime. The intention here, however, is to provide an empirical base for understanding how space and time and social characteristics are mutually modifying, interacting dimensions, which profoundly affect the nature, shape, impact and prevention of crime in urban areas. Drawing upon a series of geographically and socially focused victimization surveys and case studies undertaken in Britain throughout the 1980s, the primary focus for analysis will be women's experience of crime and threat of crime, in public space. It is beyond the scope of this chapter to do more than make passing reference to the relationship between crime against women in public *and* private space. The particular time dimension examined is restricted to crime which occurs day/night over the twelve-month period prior to the survey.

The reason for focusing on female victimization for the purposes of this chapter is twofold. First, female victimization is of especial interest to geographers and criminologists since one of the most crucial factors affecting risk of crime, fear of crime and geographical mobility in urban areas is gender (the others being age and lifestyle), and second, the data are used to challenge the conventional wisdom that crime in urban areas is predominantly a problem for men. It is to this accepted tenet that I now wish to turn.

DEFINITION AND MEASUREMENT OF FEMALE VICTIMIZATION IN URBAN AREAS: THE LIMITATIONS OF GLOBAL RISK RATES FOR MEN AND WOMEN

Conventional criminology informs us that crime is predominantly an urban, and male, problem. From Henry Mayhew (1861) to Patricia Mayhew (1989), cartographic criminology has contoured the spatial distribution of crime and located it as a problem of large towns and cities. Until recently, however, little could be said with any confidence about which individuals were most likely to be victimized or even how many people were likely to be victimized. The development of national victimization surveys in Britain throughout the 1980s has produced a more accurate estimate of the distribution and extent of crime than that provided by the official crime statistics compiled by the police. Official statistics are widely acknowledged to be an unreliable source of information about crime because of the twin problems of under-reporting and under-recording. Many victims do not report criminal incidents to the police and even crimes which are reported may subsequently be 'down crimed' or 'no crimed' by the police if they are considered insufficiently serious to merit police attention (Lea and Young 1984; Jones *et al.* 1986; Mayhew *et al.* 1989).

Among other things, victimization surveys are designed to reveal the 'dark figure' of crimes; in Britain the Home Office Research and Planning Unit financed a national programme of victimization research. Conducted in three sweeps in 1982, 1984 and 1988, the British Crime Survey (BCS) has proved valuable in estimating *general* crime trends and levels of unreported and unrecorded crime across the country, the spatial and social dimensions of risk and the impact of crime (and latterly harassment) upon victims.

The story told is now so familiar to criminological *aficionados* that not much time need be given over to it. Briefly, the BCS finds that nationally, fear of crime is more of a problem than crime itself. The 1982 survey reported that criminal victimization was a rare occurrence for the average citizen, who can expect a burglary once every 40 years; a car stolen once every 60 years; an assault once every 100 years and a robbery every 500 years (Hough and Mayhew 1983).

Secondary analysis of BCS data has provided some information as to the spatial and social concentration of crime. Spatially, crime is focused by area of residence and housing tenure with those residing in high-status, non-family and multi-racial areas of the inner city and council tenants on poorer estates experiencing increased risks of victimization (Hope and Hough 1988). It is also evident that risks of crime are differentially focused on particular individuals according to their gender, age and life-style. Thus, it is young men rather than women who are most likely to be victimized; the risks of personal (and violent) victimization being closely associated with life-style, the strongest measure being the number of evenings spent outside the home. Going out more than two evenings a week, particularly at weekends, frequenting public houses and drinking, all increase risks of street robbery and assault (Gottfredson 1984). As Gottfredson notes, the likelihood of criminal victimization from personal crime increases at night (1984: 12, 37) and that women who go out after dark in inner city areas, especially if they visit public houses at weekends, run similar risks to men, and, moreover, 'Women who go out two nights a weekend have a likelihood of personal victimisation (14 per cent) that is in excess of the likelihood of men who did not go out at all (8 per cent)' (Gottfredson 1984: 11).

The BCS emphasized the disparity between 'objective' risks and subjective fears, thereby exploding some popular myths and media stereotypes as to the frequency and focus of victimization. The surveys stressed that women and elderly people objectively experience a lower frequency of victimization though both groups express high levels of fear for their personal safety, particularly at night. With regard to women, the BCS demonstrated that their generalized fear could not be explained by the frequency of rape and sexual assault, risks which were found to be negligible. The first BCS found only one attempted rape among the 11,000 households surveyed. The 1984 BCS stressed, 'The number of sexual offences uncovered by the second survey was still very small – 19 cases among nearly 6,000 women' (Hough and Mayhew 1985: 10–11); 'Only one in forty' women mentioned incidents of sexual harassment (Hough and Mayhew 1985: 11). By 1988, despite considerable increases in the number of rapes and sexual offences recorded by the police, little appeared to have changed: 'In all, only some 15 sexual offences were uncovered by the survey among nearly 5,500 women' (Mayhew *et al.* 1989: 8).

The anomaly of women's low frequency of victimization contrasting markedly with extensive fears for their personal safety was resolved by blaming the victim. Women's fears were irrational; what was needed was more and better publicized information as to the rarity of criminal victimization in general and sexual offences in particular. Gottfredson (1984) summarized the Home Office position:

> One of the things that surveys like the BCS can do, is inform both the public and policy makers about the distribution of criminal victimisation. One aspect which stands out in the BCS and in every victimisation survey to date is the relative rarity of serious criminal victimisation. ... The data contained herein can help the public to know that crime is relatively rare, that most crime which does occur is not particularly serious in the sense of bodily harm and financial loss and that a small segment of the population seems particularly prone ... it is the young who are most at risk.
>
> (Gottfredson 1984: 30)

Quite so. Though secondary analysis of BCS data provides some information as to spatial and social factors associated with victimization, the analyses presented are (as the BCS authors acknowledge) extremely limited. Spatial variations based on ACORN classification do not refer to specific areas of town or cities but to different *types* of areas across all towns and cities (Hope and Hough 1988) and as small area surveys have noted, even in the confined space of the inner city, there are marked social variations in the experiences of crime, between and *within* different subgroups of the population.

The main criticism that has been levelled at the BCS is that global statistics which aggregate high and low crime rates in different parts of the country mask the geographical and social concentration of crime at an intra-urban level (Young 1988) and, as Smith (1989b) observes, despite all of the political rhetoric and the politicization of Law and Order problems throughout the 1980s we actually know very little about the true extent of inter- and intra-urban crime. What is becoming increasingly clear is that however much of a problem crime is in urban areas, not only is it focused on particular individuals in time and space but also the impact of crime varies enormously in relation to the differential, social and physical vulnerability of those victimized (Maguire 1982; Lea and Young 1984).

The specific topic of women's experiences and fears of crime highlighted further shortcomings of national victimization research. The BCS has been criticized for recording and concentrating on women's fear of crime to the exclusion of their everyday, common experience of interpersonal abuse in public places which greatly contributes to women's fear of assault (Stanko 1987). By concentrating on objective, discrete legal crime categories, the BCS glossed over vital material differences between victims and the focus on the quantity of victimization rather than the quality of it ignored the subjective aspects of victimization. There is no such thing as an equal victim and there are substantial differences between individuals in their capacity and tolerance to withstand crime and the threat of crime. Women constitute a glaring example of the un-equal victim because they are socially and physically vulnerable to victimization (Stanko 1985; Young 1988). Women experience a range of offensive behaviour directed at their sexuality, which they may perceive as victimization but which would not necessarily be deemed 'criminal' which, nevertheless, profoundly shapes women's lives creating a very different social reality for women than for men. The tendency for macro-level quantitative surveys to adopt a definition of crime independent of the subjective definitions of victims meant in Young's words, that 'Victimisation research commonly trivialises that which is important and makes important that which is trivial' (Young 1988: 173). Additionally, it has been pointed out that crime surveys also have an inbuilt 'dark figure' in relation to crimes against women. It is now generally accepted that women and girls are involved in hidden and unreported violence which occurs in private household space and that large-scale victimization surveys are not the most appropriate instrument for uncovering the true extent of family violence (Hanmer and Saunders 1984; Kelly 1987; Stanko 1988).

The primary function of macro-level crime surveys remains to provide a comprehensive picture of the extent and distribution of reported and unreported crime across the country. In so doing they are extremely valuable because they enhance our understanding of crime and policing and the impact of both on victims. Taken in isolation, though, macro surveys are empirically and theoretically incomplete. The extent and impact of victimization on individuals and communities is not something which can be adequately comprehended through empirical regularity (even if the samples are large enough to include enough individuals, to be statistically representative of the whole country) and statistical analysis can never be the sole criterion for leading the direction of criminological inquiry which in reality is grounded in the different problems, specific people face in the real world.

SMALL AREA STUDIES: THE SECOND GENERATION OF VICTIMIZATION SURVEYS IN THE INNER CITY

In the latter half of the 1980s the second generation victimization surveys were, and continue to be, developed in the United Kingdom. They represent an attempt to refine both the geographical and social dimension of risk, in parallel. By concentrating on small areas within the inner city they are better able to define areas of high crime. By high sampling of populations they offer a detailed social focus by gender, ethnicity and age and they illustrate how these social factors combine with each other to heighten or reduce the probability of victimization in relation to the social structure of the locality. Some of these surveys have used maps to pinpoint the incidence of crime and avoidance of physical space (Lea *et al.* 1988) and others refine the focus further, mapping victimization by locality, gender and time (Painter 1988; 1989a; 1989b). Local inner city surveys aim to go beyond the abstraction of risk rates based on aggregate statistics for England and Wales as a whole and have emphasized the need for criminology to address the specific, the particular, the singularity of people/place crime problems (Forrester *et al.* 1988). Moreover, they have not simply adopted a level of analysis in a way which could be marginalized by mainstream criminology as 'micro level', or 'atypical' since they have a strong empirical basis in time, space and number. It may be useful, therefore, to explain how these surveys have evolved.

Because crime is geographically and socially focused, recent victimization surveys have concentrated more on the micro-level of local authority boroughs, estates and streets and, in the case of the London Borough of Hammersmith and Fulham, all three.[1] Local surveys give a much more informed picture about the geographic and social focus of intra-urban crime and policing. The second generation victimization surveys have emphasized the importance of analysing spatial, social and temporal factors as interdependent, rather than independent, variables in the geography of crime and policing. It is now clear that the life paths traced by different subgroups within geographically confined areas cannot be reduced to either spatial, social or temporal factors; it is necessary to analyse all three in order to comprehend the risks and impact of crime and policing problems that people experience in their lives. For example, both the Islington and the Hammersmith and Fulham Crime Surveys demonstrated the highly focused nature of police contacts with particular subgroups. In both areas it was rare to find one middle-aged, white man or woman who had been stopped by the police yet approximatey half of all young black men had been stopped at least once in the previous twelve months. But it also became evident that differential experience within relatively small geographical areas based on race was much more complicated than the abstraction that black people are more likely to be stopped than white people. In the Hammersmith and Fulham (Painter *et al.* 1989a) and Islington (Crawford *et al.* 1990) borough-wide surveys and in the more focused estate survey in Fulham (Painter *et al.* 1989b) no black women over the age of 45 had been stopped whereas in Islington young, white women were over three times more likely to be stopped than older black men. And (as will be pointed out subsequently) in relation to women's experience of crime and harassment there is considerable variation in women's public encounters within the same geographical areas when class is introduced into the equation. Whereas geographers have a tendency to reduce analysis of problems to the spatial and sociologists have a predisposition to reduce everything down to the social, it is clear that both aspects are needed. To put it succinctly, in order to comprehend the enormous variation in urban dwellers' experience of crime and policing one has to be acquainted with both the spatial and social context of encounters. One needs to know *who* is passing through *which* space, *when*.

Taking as a baseline, BCS methods and many of its questions, recent victimization surveys have expanded the legal crime categories which were adopted by the BCS in order to compare unreported crime with that reported to, and recorded by, the police. They take a more consumer orientated, subjective and evaluative definition of what counts as serious and non-serious crime from the victims' perspective. In recognition that direct experience of criminal victimization is only one dimension of victimization and fear of crime, the surveys widened the focus of inquiry from crime, fear of crime and policing to a range of other issues including racial and sexual harassment (topics of inquiry subsequently taken up by the BCS), multi-agency intervention and council service delivery. This approach was developed in order to contextualize the problems of crime and disorder on communities and individuals within them. Repeated exposure to interpersonal abuse, disorder incivility, individual differences and differential vulnerability to the threat of crime, the real and perceived lack of recourse to avenues of redress and restitution through the criminal justice system and local political and community organizations all affect risks and fears of crime. The focus on the lived reality of differential social experience is one of the hallmarks of what has been labelled 'realist' criminology. The emphasis has been on the importance on specificity; that risks and fears of crime cannot be read off average risk rates and average citizens but must be interpreted in relation to specific social categories at specific times in specific places. The method has been to map the problems of crime and threat of crime and then put the problems in context by giving due weight to victims' perceptions of their situation and this includes a genuine appreciation that any definition of victimization must take account of significant differences in experience, evaluation as to seriousness of offence and tolerance of social behaviour.

THE EXTENT OF CRIME IN INNER CITY AREAS

Localized surveys reveal a much higher rate of victimization than BCS estimates for inner city areas and council estates located in the inner and outer city. The Hammersmith and Fulham Crime and Policing survey carried out in West London in 1988 encompassing 1,315 households (Painter *et al.* 1989a) showed that one out of every two households had been touched by serious crime within the previous twelve months; a figure replicated by a population study of 452 households on an estate within the same borough, using the same questionnaire (Painter *et al.* 1989b) and in North London by the second Islington Crime Survey in which over 1,600 interviews were completed (Crawford *et al.* 1990). To the extent that there is any pattern discernible it appears that crime is becoming more, not less of a problem. It is undoubtedly a frequent occurrence in urban areas and women are not exempt.

THE SPECTRUM OF CRIME AGAINST WOMEN IN PUBLIC SPACE

Local victim surveys will undoubtedly underestimate the amount of crime against women but they do provide an additional source of empirical information and elucidation of the extent, nature and impact of unrecorded instances of crime and threats against women. In particular they have shown 'objectively', in pockets of the inner city and on peripheral council estates, that

1 women are, proportionally, more likely to be the victims of crime than men
2 women experience a wider spectrum of crime and threat of crime than men
3 women have particular crimes and threats focused upon them.

Realist criminology has been concerned to produce objective measures of crime as well as subjective evaluations as to what constitutes serious and non-serious crime from the victim's perspective. Bearing in mind that victims' definitions will not necessarily correspond to legal categories used by the police and Home Office one is immediately confronted with a conceptual problem of definition and measurement. As mentioned above, the second generation of victimization surveys have extended their conceptual and empirical range of inquiry to include activities which constitute victimization. But they are behaviours which do not fit easily into official crime categories and are, therefore, not legally recognized as crimes.

But what do local surveys tell us about crimes against women? Despite prevailing conceptions that women are relatively rarely the victims of violent personal offences, the first Islington Crime Survey (ICS1) found that women were 40 per cent more likely to be a victim of street robbery than men and twice as likely to be assaulted and assaults were more likely to involve violence causing physical injury (Jones *et al.* 1986). The trend within the borough was confirmed by subsequent surveys. The Hilldrop Estate Survey (Lea *et al.* 1988) found that combining all personal criminal offences women had a higher rate of victimization (8 per cent) than men (5 per cent). The second sweep of the Islington Survey revealed yet again that women were twice as likely (8.5 per cent) as men (4 per cent) to be a victim of street robbery and personal theft in a public place (Crawford *et al.* 1990).

The surveys also revealed a much higher incidence of sexual assault than the BCS estimates. ICS1 established that sexual assault in Islington was fourteen times higher than the BCS average and that sexual assault was focused on young women aged 16 to 24 who were eighteen times more likely to be sexually assaulted than women aged over 45 (Jones *et al.* 1986). ICS2 disclosed that within the previous twelve months one out of fifty women in Islington had been sexually assaulted. The inclusion of questions on sexual abuse which asked about experiences over a lifetime (rather than the usual format 'During the last twelve months ...') found that one out of thirteen girls under the age of 16 had been sexually abused and that women were twice as likely (1:100) as men (1:200) to have been the victims of intra-familial abuse (incest).

And there was nothing peculiar or exceptional about Islington (at least not in respect of crime). The Hammersmith and Fulham Borough Survey (Painter *et al.* 1989a) found that women had an equal chance to men of falling victim to street robbery and physical assault; a finding replicated to a similar degree by a local estate survey conducted in the same geographical area and using the same questionnaire (see Table 9.1; also Painter *et al.* 1989b). So even if one just examines the incidence of crime, as legally defined, the disparity between women's low risk and high fear as measured by macro-surveys, disappears. Not only do city women objectively run equal risk of personal crime, but also they experience a wider spectrum of crime than men, sexual offences and harassment being almost exclusively a female entitlement. The extent of sexual harassment or 'non-criminal street violence' (Jones et al. 1986) has also been documented by recent victimization surveys and it is clear that this type of behaviour is a commonplace and inhibiting feature of women's lives, restricting their access to and freedom of movement within public space. It does appear from the data that women experience greater levels of threatening and abusive behaviour in public places; much of it directed at their sexuality. It is, as Table 9.1 shows, a much greater problem in some areas than others: twice as many women interviewed on the estate stated that they had experienced sexual harassment and six times as many women than men on the estate and twice as many women as men across the borough experienced general harassment.[2]

Both the borough survey and estate survey conducted in Hammersmith and Fulham revealed that twice as many women had been harassed than men and over four times as many women have been multiply harassed. One in nine women across the borough had been sexually harassed, pestered and threatened (a figure four times higher than the BCS average of one in forty) with half of these reporting that this type of incident had occurred between three and six times (and more) within the previous twelve months.

The detailed focus on subgroups within a well-defined area illustrated that within a specific subgroup of the population there are significant variations in women's experience of harassment. Tables 9.2 and 9.3 show that within the Borough of Hammersmith and Fulham, the most harassed group are young women under 25, almost 60 per cent of whom have been harassed: the lowest are men over 45, the difference between men and women being a factor of

Table 9.1 Proportions of men and women experiencing crime and harassment

	Borough Survey (N = 1,315)		Estate Survey (N = 452)		Total	
	Men	Women	Men	Women	Men	Women
Physical assault	6	3	4	7	10	10
Street robbery	6	6	5	5	11	11
Sexual assault	–	1	–	1	–	2
General harassment	19	41	3	18	22	59
Sexual harassment	–	7	–	14	–	21

Source: Adapted from Painter et al. 1989b

six. Table 9.2 outlines the wide disparities in experience of harassment when age and gender are combined; note the extremely high rate of repeated harassment of young women, one in five of whom had experienced three or more incidents within the previous twelve months. Table 9.3 records the incidence of specific acts of harassment by gender.

Table 9.2 Level of harassment by combination of age and gender (%)

	No harassment	One or two incidents	Three or more incidents
Men			
16–24	73	23	4
25–44	79	16	4
45 and over	91	9	0.5
Women			
16–24	41	38	21
25–44	56	29	15
45 and over	83	16	1

Source: Painter *et al.* 1989a: 18

Table 9.3 Annoying and frightening behaviour by gender

	Men	*Women*
Stared at	8	17
Followed	5	20
Approached or spoken to	5	19
Shouted at or called after	10	20
Touched or held	3	10
Kerb crawled	2	8

Source: Painter *et al.* 1989a: 18

Crime, harassment and the constant threat of crime in urban areas create a very different social and spatial reality for women in the inner city. Sexual harassment is a common experience for a significant number of women, particularly young women, and it creates insecurity and wariness of open spaces. Women are particularly vulnerable to incidents of sexual harassment since the unknown outcome of being approached, followed, shouted at or molested may well be the precursor to a more serious assault, particularly if it occurs when a woman is out alone after dark. Some commentators have argued that hostile and abusive behaviour to women triggers their fear of rape and sexual assault (Warr 1985) and death (McNeill

1987). Certainly, we now have a substantial body of research which shows it is women who are primarily the victims of public abuse and harassment and that fear increases with the incidence of abuse, reinforcing its importance as a factor in fear of crime. The second Islington Crime Survey found that 43 per cent of all women interviewed had experienced some form of harassment in public places within the previous year and, in line with the trends outlined above, that young women were at greatest risk. The survey also demonstrated the relationship between public abuse of women and their willingness to use the streets after dark. A staggering 87 per cent of women who had been victimized at least once, and 97 per cent of those who had been victimized on numerous occasions said that they were unwilling to use the streets at night (Crawford *et al.* 1990).

One has only to take into account that sexual violence is an ever-present threat which does not affect men and add to that the greater amount of harassment and crime they experience in public and private settings, to understand women's fear of crime as a rational and realistic assessment of their physical and social vulnerability. And the fact that fear of crime is not confined to those women on the survey who have directly experienced harassment does not mean that other women's fear is exaggerated or irrational. Victim surveys traditionally confine questions to a time limit of the previous twelve months and it is evident that within this time span young women have greater direct experience of abusive behaviour. But all women have been young at some time and so within a lifetime's experience will have encountered abusive and frightening behaviour. There is also evidence that witnessing incidents of crime and harassment also contributes to levels of fear; it is to be remembered that most of the incidents that surveys pick up occur in public places and are therefore likely to be witnessed by other women, which may well continue to reinforce women's vulnerability whether or not they have directly experienced incidents within the previous year or more.

It is therefore no exaggeration to assert that sexual harassment is one of the most effective mechanisms of social control in that it perpetually reminds unaccompanied women that they are not welcome, or safe, in public space. The frequency and impact of harassment on women's lives illustrates the false distinction which is often drawn between serious and non-serious crime and in differentiating female victims from non-victims. Crime and harassment exist on the same continuum of violence towards women: a continuum which should not be seen as a linear line from serious to non-serious crime as legally constructed, but more as a series of layers and slices of life experiences which make up a lifetime biography of subordination, social control and oppression. Women's experiences of violence and sexual threat are lifetime experiences which will not be reflected in survey questions which ask about experiences over the previous twelve months. Repeated exposure to threatening situations throughout a lifetime means that the threat of crime merges into and out of the reality of crime and over time. This creates a hostile and intimidating public environment which is reinforced on each occasion of abuse; and the threat is posed not by individual men, but by men generally.

To put it bluntly, the women surveyed, do not fear crime they fear men and it is a real fear; fear which limits their freedom of movement, where they can go, when they can go, how they can go and with whom they can go. Fear which alters their perceptions of space and the built environment (Painter 1988; 1989a; 1989b); fear which reminds them that men dominate public space and control access to it.

THE IMPACT OF VICTIMIZATION ON FEAR OF CRIME AND WOMEN'S USE OF PUBLIC SPACE

An important component of local victimization surveys has been to chart how victimization and fear govern women's use of the urban environment and, in so doing, they further illuminate connections between spatial structures and patriarchal social relations. In deconstructing women's fear, local surveys have shown that the relationship between female victimization and fear increases as the type of crime is more closely associated with the space that the fear governs. Accordingly, victims of crime in the streets fear public spaces at night more than non-victims; women who have been sexually or physically assaulted in public or private space express higher levels of fear in relation to all crime.

A consistent finding of all victimization data is the extensive precautionary behaviour adopted by women in urban areas to reduce their risk of victimization from crime and harassment. The majority of city women avoid being alone in public space, they avoid public transport, subways and they rely on men as escorts particularly after dark (Hough and Mayhew 1983; 1985; Jones *et al.* 1986; Painter *et al.* 1989a; 1989b; 1990; Crawford *et al.* 1990). Table

Table 9.4 Proportions of men and women undertaking avoidance strategies
to negotiate public space in inner city areas

Avoidance strategy	Hamm and Fulham Borough Survey (N = 1,315)		Islington 2 Borough Survey (N = 1,621)		West Ken Estate (N = 452)		Ladywood Estate (N = 514)	
	Men	Women	Men	Women	Men	Women	Men	Women
Avoid going out after dark	23	65	39	71	39	68	48	71
Avoid certain streets	55	78	17	43	53	72	56	71
Go out with someone rather than alone	60	89	17	46	17	59	17	60

Source: Painter *et al.* 1990: 42

9.4 charts some of these factors across recent estate and borough-wide surveys.

It is all too evident that when night falls, the majority of the women surveyed, live under a self-imposed curfew; although Table 9.4 shows that men also undertake some precautions to reduce their risks of victimization, across all areas and strategies the figures for women are much higher. Women's avoidance of public space after dark must, however, be interpreted in a broader social context. Precautionary behaviour is practised in large part to reduce women's exposure to victimization from crime and the threat of crime, but it also reflects other aspects of patriarchal oppression which confine women to the home and deny them opportunities and access to leisure and relaxation after working either inside or outside the home during the day. Leisure opportunities are manifestly mediated by class and culture but as a general rule women as a subgroup do not engage in leisure activities to the same extent as men. In fact they have been officially designated by the Department of the Environment as recreationally 'disadvantaged', and by the Sports Council and Social Science Research Council Joint Working Party on Recreational Research, as 'socially and geographically deprived' (quoted in Green *et al.* 1987).

Notwithstanding the general finding that female victimization from crime and the threat of crime is widespread it is clearly a misfortune which is suffered unevenly within the subgroup.

Contemporary victimization surveys illustrate the fallacy of reducing urban experience of victimization solely to any one factor such as gender, race, age or class even within small geographical areas. Such reductionism obscures the wide variations in social and economic conditions within subgroups which are structured in terms of all these constituents. The specific social focus of small area victimization surveys have thus highlighted important connections between the geography of social space and wider social relations, in the victimization of women.

The specificity inherent in the second-generation crime surveys does accentuate the differences between women as a social group. Being 'a woman' is not a crystalline, exclusive, unambiguous social category; it is mediated by class, age and ethnicity. All western women are born into two sets of dominant material relationships – patriarchy and class. Patriarchy renders all women unequal; class position means that some are born more unequal than others. Differentials in wealth affect the manner in which different women (and men) live in, and move through, public space thereby influencing the pressures and risks of crime in their everyday encounters. Middle-class women can minimize their risk of crime and harassment in public space by using a car rather than walking or using public transport. And, it is to be borne in mind that modes of transport do affect likelihood of victimization, with those reliant on underground/train having twice the rate of victimization (26 per cent) than those using cars (10 per cent) (Gottfredson 1984: 10). Access to a household car provides safe transport through space, and the leisure and social venues associated with middle-class lifestyle are more secure and protected. Middle-class women have access to additional avoidance strategies which not only limit their risk of victimization but also mean that they may routinely use different time-space paths across wider areas of the city, thereby exercising more control over their use of public areas over longer periods of time. In a nutshell, they are more able to move freely in time and space because they are better protected; the typical pattern of movement for a young, middle-class, professional woman can be summed up as movement from house, by car, to wine bar, theatre or friend's house, returning by car or taxi to home. (And for the cynic who may comment that middle-class women have a greater chance of having their car stolen, they may be inconvenienced but they will also be insured.)

Greater wealth increases access to a range of urban facilities and weakens geographical constraints for some women as to the type of area and dwelling they inhabit and their means of movement through time and space. It means that some women rarely confront the reality or threat of victimization in their everyday interactions. Of course the fact that middle-class women exercise any constraint at all emphasizes that it is the social context of womanness that is the root cause of male to female victimization. Yet, the life experience of middle-class women can render them less emotionally and physically vulnerable to crime, threat of crime and fear of crime than say working-class women living in the same area. For example, on a recent survey of a deprived inner city estate in Birmingham, seven out of ten women interviewed, who described themselves as working class or poor, did not have access to a car, leisure facilities or employment opportunities and as a consequence of combined factors, including a generally high level of criminal victimization on the estate, seldom ventured out after dark (Painter *et al.* 1990).

Thus, if one wanted to estimate or predict the incidence and impact of female victimization upon access to urban facilities and public space, one could not do this by abstracting risk solely from any single social category such as gender, or from spatial factors such as living in the 'inner' or 'outer' city. It is important to analyse risk rates in terms of both factors simultaenously. Working-class women living in the outer reaches of the city and upon peripheral council estates may experience extraordinary high crime rates (Painter 1988) whereas middle-class women living in the inner city may experience very little crime (Crawford *et al.* 1990). Some women on an estate will be subject to almost continuous harassment, others will not (Painter *et al.* 1989b) and evidence from the Islington Crime Surveys (Jones *et al.* 1986; Crawford *et al.* 1990) and a recent survey of marital rape (Painter 1990) indicates that working-class women vulnerable to personal victimization in public space are also more at risk in private space.

A realist approach emphasizes the need to be aware of the specificity of generalization and the need to base analysis of victimization in specific areas and social groupings. Although realist crime surveys analyse interactions between people in confined space and time, they also recognize the need to connect specific spatial and social relations to wider structural, ideological and social processes, which provide, and strongly influence, the social and economic context of specific environments and are the root cause of antisocial, face-to-face interactions between female victims and male offenders.

CRIME PREVENTION AND FEAR REDUCTION IN PUBLIC SPACE WITH SPECIAL FOCUS ON WOMEN

Avoidance of public places may reduce women's exposure to risk of victimization but there are costs to communities generally, and women in particular. Avoidance tactics further increase and compound women's social isolation by limiting their access to public space and making them more dependent on private household space where (as will be demonstrated) violence may also be located.

Withdrawal from social and leisure opportunities also affects the commission and prevention of crime and the threat of crime. It results in streets that are empty of people, a reduction in the possibility of human intervention should crime occur and the absence of possible witnesses to incidents which are the primary aid to the detection of crime. In other words, avoidance of the public sphere undermines a range of informal community support mechanisms that, if in place, may well assist in crime prevention, fear reduction and victim support in the event of victimization. The early work of Jacobs (1961) and Newman (1972) was instrumental in focusing attention upon the physical environment as a conditioner of offence behaviour and in particular in highlighting the relationship between the design and layout of the space within which crime occurs and the manner in which localities are informally controlled and observed. In essence, they contended that areas are more likely to be informally controlled if they are visible to potential witnesses, if the design is such that a continuous stream of potential witnesses is encouraged and if social support networks are encouraged so that residents and pedestrians are encouraged to protect neutral and publicly used territory.

Women's space evasion at night does not tend to increase personal safety so much as compounding fear and undermining the quality of life in urban areas. Furthermore, there is now a body of research which has shown that the withdrawal of social support networks exacerbates fear of crime by undermining individual feelings of personal safety and autonomy and this in turn contributes to a belief that the lived and built environment is uncontrolled, unpredictable and crime-prone (Lewis and Salem 1981; Taylor *et al.* 1986; Smith 1989a). The design and physical features of urban areas have a definite influence on social psychology of the lived environment. Signs of physical deterioration including graffiti, vandalized and unrepaired property, litter, poor lighting, combined with disorderly behaviour such as youths loitering, noisy neighbourly disputes and verbal abuse generate fear because they signal an environment that is out of control, inhabited by unpredictable and belligerent people (Lewis and Maxfield 1980; Skogan 1987).

Because of their unequal physical and social vulnerability, women feel particularly threatened by obvious signs of physical deterioration and social disorder and the prospective threat increases at night to the extent that their use of public space is severely and habitually curtailed. The second Islington Crime Survey specifically examined the proposition that environmental conditions contributed to fear of crime in the street. The survey looked at three items which pose the most obvious indicators of street disorder and deterioration: poor street lighting, vandalism and dirty streets. It found a direct causal relationship between fear and negative attitude towards physical aspects of the environment across all social variables. Almost half the women (47 per cent) who expressed high levels of fear also rated the three items as 'big' problems in their area (Crawford et al. 1990: 83).

There is now ample evidence that not only do environmental conditions affect fear of crime, but also they affect opportunities for crime and harassment to occur and may well influence offenders' perceptions as to the likely risks of being apprehended. In particular, Mayhew et al. (1979) highlighted the importance of surveillance as a deterrent to crime in that offenders usually try not to be seen when committing criminal acts and that offenders do appear to be deterred by the actual or potential presence of the public.

Surveillance, by its very nature, means visibility. Recently a series of exploratory, micro-environmental projects into the effects of street lighting on 'outside' night-time crime, fear of crime, personal safety and pedestrian traffic flow have provided an interesting insight into the routinized gender differentiated time-space paths traced by individuals as twilight descends. They also demonstrate how space and social events which occur within space are perceived differently and take on a different significance after dark. In so doing, the programme provides further insights into connections between the design, use and form of the built environment and the victimization of women as a reflection of what Giddens has called the 'much broader properties of the institutionalisation of social life' (Giddens 1985: 272).

The lighting and crime prevention research tapped into the experiences and perceptions of women who being out at night, were out of role and out of context. Gender inequalities in British society mean that it is not customary, or indeed natural, for women to occupy public space, particularly if unaccompanied and especially after dark. This is a social law which women learn early. Traditionally their activities are limited to the private space of home and family which form the principal sites of informal social control and keep women in their place. Even in the 1990s it is still not socially acceptable, let alone safe, for a woman to walk the streets alone at night. If a woman transgresses this social norm by going into public places, particularly public houses, alone:

> She is seen to have given up her entitlement to the protection from any man. ... Such women are assumed to be 'available' and are expected to be receptive to sexual advances from men ... [and] are deemed to be responsible for their own behaviour, and for the behaviour of the men they come into contact with. Police comments upon, and media coverage of, reported sexual attacks highlight the distinction between legitimate and illegitimate violence against women. They frequently focus upon the *time and place* where the attack took place, with women being seen as particularly culpable if they are unescorted in public places after dark.
>
> (Green *et al.* 1987: 86; 89)

The research programme was designed to evaluate the scope for improved street lighting as a means of crime prevention under spatially and socially controlled conditions. Men and women pedestrians were asked about their experiences of 'outside' crime and nuisance, their fear of crime and perceived risks of personal safety when using badly lit streets. These variables were compared before and after improvements were made to public lighting and some interesting insights into women's victimization, urban life-style and their perceptions and usage of the built environment emerged.

The lighting and crime prevention studies adopted on-street surveys and the majority of interviews took place after nightfall. They demonstrated that women who do go out after dark have an equal if not greater risk of victimization from crime and harassment (Table 9.5). They also revealed that women who do go out are not less fearful than women who stay at home but a significant number have to take more elaborate precautions in order to negotiate public space at night and reduce their risks of victimization. Typical precautions took the form of 'street nous', that is walking in the road rather than on the pavement, never carrying a handbag, dressing 'down', wearing flat shoes to facilitate possibility of escape, running rather than walking, limiting the amount of money carried and jewellery worn, going out only if accompanied by a safe man or a large dog. One of the clearest indictments on the appalling quality of urban life was the finding that approximately one in three women interviewed prepared for the eventuality of attack before leaving home and carried an object or alarm to protect themselves in the event of victimization. Table 9.5 outlines victimization rates for men and women in poorly lit areas after dark and it is important to note that victimization of women recorded in the micro-environmental surveys occurs in the context of women taking considerably greater precautions to avoid victimization than men. Women's constant unease at night resulted in a greater sensitivity to, and awareness of, physical and social aspects of the built environment. It appeared that women were continuously alert to the possibility of danger and therefore consciously and unconsciously monitored their surroundings for safe and unsafe places to walk. No doubt because of this, women were more likely than men to notice improvemens in street lighting – improvements which increased their confidence and sense of personal safety at night while using the street.

Table 9.5 Victimization rates of men and women in poorly lit areas after dark

	Hammersmith		Edmonton		Tower Hamlets		Total	
	Men	Women	Men	Women	Men	Women	Men	Women
Physical attack	9	14	9	9	4	0	22	23
Threats/pester	8	25	18	24	20	20	46	69

Source: Painter *et al.* 1989b: 75

Women's fear of crime in all three areas was reduced by altering one facet of the physical environment. But this did not occur simply because recognition of people and place were enhanced by superior lighting levels. The improved public lighting also changed local social relations and social usage of the streets. Incidents of crime and harassment dropped in all three areas and pedestrian traffic flow increased. In one of the areas, groups of youths ceased to congregate in the street and men using a neary public house stopped urinating in the street and surrounding gardens. The reduction in crime and nuisance combined with good night-time visibility positively affected perceptions of crime in the immediate area. Across all three projects the majority of pedestrians believed that street robbery, physical and sexual assault, and threatening behaviour had decreased.

Overall, the impact of good lighting was much greater on women than men and the task of accounting for this differential experience remains a matter for conjecture given the lack of research into women's experience of urban life, social change and the built environment. Nevertheless a number of interdependent physical and social processes appear to be at work in reducing women's anxieties concerning their actual and perceived risk of victimization at different times.

First, environmental and spatial structures and the meaning attached to them change at night. Darkness physically alters space. Poor lighting levels and/or poorly maintained public lighting proffer one of the most obvious signs of environmental degradation at night, creating in many parts of the city an impression of a hostile, uncontrolled and therefore unsafe environment. The psychology of this relationship has already been referred to, in that signs of environmental deterioration result in a loss of autonomy and personal control over space, which in turn affects social behaviour. Good lighting may well be a signal to women that they are traversing a well-ordered and controlled environment within which they are less likely to be attacked and within which the possibility of human intervention in the event of victimization is a real possibility.

Movement through space is also movement through time (and this does mean more than saying that it takes time to go from *A* to *B*). Space is not simply demarcated by physical boundaries; it is also marked out in time and night-time is one of the most obvious time space delimitations affecting social interaction. Consequently, it seems reasonable to conjecture that interactions between female victims and male offenders take on a different significance at different times and in different spatial contexts. Women are subject to harassment in public places during daylight and darkness. The reaction to this will vary according to individual and spatial factors but whereas verbal abuse, flashing, unwanted touching and being followed during the day may be shrugged off, laughed at or be challenged by a woman, at night, the same behaviour would be regarded as more threatening.

Men and women are more apprehensive of victimization in darkness than in daylight, a phenomenon which Walsh (1983) puts down to an invocation of childhood fears of ghosts and strangers. Because of their unique vulnerability to sexual assault, women are

185

more insecure and fearful; though good public lighting will not eradicate the causes nor incidence of sexual violence towards women, it does seem that it has influenced the potentiality for female victimization at night by altering the spatial structures and affecting social relations within them. Because not enough is known about the risks and consequences men attach to committing violent and abusive acts towards women not much more can be said about the direct cause and effect relationship between human and environmental factors as they move through time. Nevertheless, as Mayhew *et al.* (1979) have noted, it is a fact that offenders usually try to avoid being seen. Consequently, it seems reasonable to conjecture that offenders place more weight on the immediate risks of being recognized and challenged by members of the public than they do to the serious, but remote, consequences of apprehension by the police and conviction by the courts. If this is the case, then the manipulation of offenders' fears of being seen and challenged at night can be manipulated by well-lit environments in the interest of preventing crime and fear among women.

PUBLIC VERSUS PRIVATIZED CRIME PREVENTION INITIATIVES

By producing evidence of the connections between fear of attack in public places, the detrimental effects of avoidance behaviour on individuals and neighbourhoods at night, social disorder and environmental deterioration, surveys have drawn attention to the possibilities of controlling crime and reducing fear through a variety of urban policies and targeted strategies via investment in the fabric and facilities of local areas. Though local government have undertaken many initiatives in relation to crime prevention and community safety, and some have recognized the specific needs of women in these respects, central government continues to push for privatized rather than public crime prevention strategies.

Throughout the 1980s individuals have been exhorted to take responsibility for crime prevention and it does make sense that those who can afford to do so take precautions themselves, for example by installing household security. But those who are at greatest risk of crime frequently reside in the poorest public sector housing and can ill afford household security and insurance. The increasing pressure on local authorities to cut costs and privatize essential services has resulted over time in a lack of investment in, and maintenance of, the physical conditions of the built environment in many areas. These are the areas where the litter collects, the lighting is old, badly maintained or vandalized; where the offence-prone environment develops, where youths congregate to hassle and hussle; where citizens are anxious about the lack of control they are able to exert over their environment. These are the areas where the 'flight or fight' syndrome operates. Those who are able, or have to, take to the streets prepared to defend themselves from the possibility of attack; the rest either move away or retreat into the privacy of their homes and where women in particular live under self-imposed house arrest. It is, after all, in these areas that crime problems are greatest and where individuals are taking individual responsibility for their own personal security. They are doing so by

1 never or seldom going out after dark because of crime
2 taking weapons, alarms or large dogs for individual protection
3 restricting movements to narrowly circumscribed routes, avoiding streets and public transport in their communities.

The cumulative effect of these individualized strategies and lack of investment in local services is to clear public places of people, encourage a siege mentality, increase opportunities for crime and incivilities to occur and contribute to the spiralling downward trend of urban degeneration and the fragmentation of community safety. And all of these processes impinge more on women than on men. Crime prevention literature advises people to take greater control of their environment but this can succeed only if individual efforts are reinforced by public investment in the services and physical structures of neighbourhoods and the special needs of subgroups, such as women, who inhabit them.

The cause of the vast majority of crime against women lies in patriarchal social relations which extend way beyond the geographical contours of neighbourhoods and localities. As long as our society continues to define 'a real man' as someone who always get what he wants whatever the costs and consequences and the ideal woman as someone who is passive or submissive, someone who puts up, and ideally shuts up, then the causes of violent crime against women will remain unaltered. But given that feminist politics has a lot to do before inequalities in power relations between the sexes is achieved, it is vital that the needs of women are addressed through a variety of urban policies. Of particular concern is the victimization of women in the domestic sphere. Thus far, discussion has centred on victimization in public space and women's attempts to alter their life paths in time and space in order to avoid it. As mentioned earlier, there are real costs and limits to these strategies not least because female victimization is not restricted to public locations but occurs within private household space. In the 1970s and 1980s there has been a notable increase in feminist literature and research into violence against women which occurs in the home. The increased public awareness of the susceptibility of women and girls to physical and sexual abuse has coincided with the development of this literature.

WOMEN AT RISK: THE LINK BETWEEN PUBLIC AND PRIVATE SPHERES AND THE MEANING ATTACHED TO HOUSEHOLD SPACE

Men are violent to women because they know, for the most part, that they can get away with it and the issue of violent crime against women illustrates that there is in fact no sharp division between public and private space. A recent national survey, the first to focus exclusively on wife rape shows that the problem of physical and sexual abuse is widespread (Painter 1990). Out of a representative sample of 1,007 married women, one in seven (140) had been raped by their husbands (that is, if the legal definition as 'sexual intercourse without consent' applied to them) and almost half the rapes (44 per cent) involved threatened or actual violence. Apart from these experiences, one in three women surveyed, reported being threatened or hit by their husbands. The research indicates how physical space and social relations which occur within it, carry gender-based meanings which affect the interpretation of violent acts. Although the majority of all women surveyed thought that rape by a husband was equally as serious as rape by a stranger or other acquaintance, women actually raped by their husbands were likely to downgrade their experiences. Rather than being prone to 'cry rape', the majority of women who were raped, even those raped with violence, did not define the event as 'rape' at the time it occurred. The social and emotional relationship between victim and offender, the meaning attached to the space within which the abuse takes place and the cultural myths and pressures supporting the idealized family home as a 'safe haven in a heartless world', make it difficult for women not only to talk about their experiences of violence, but also to recognize their violation for what it is.

Minimizing the importance of being assaulted by a man with whom a women has a relationship is not unusual (Hanmer and Saunders 1984) but the 'wife rape' survey does lend weight to a conjectural hypothesis that women who are vulnerable to violence and abuse from people they know in private space are also at greater risk from strangers in public space. For example, working-class women in the marital rape survey were twice as likely to be raped and three times more likely to be raped with violence than women higher up the social spectrum; they were five times more likely to be raped by a stranger and four times more likely to be raped by an acquaintance. Add to this, the findings of localized crime surveys (Jones *et al.* 1986; Crawford *et al.* 1990), which show that working-class women also run higher risks of crime in public space and it is pretty convincing to argue that the root cause of crime against women is patriarchy, a social relationship which permeates both the public and private domain. What seems to emerge from the cumulative findings of recent surveys is that social class, geographical area and female victimization are strongly correlated and that a substantial proportion of working-class women in urban areas are more vulnerable to personal crime from men, known and unknown, in public *and* private space.

The physical location, timing and frequency of incidents all affect people's assessment of their ability to exercise control of life circumstances. And when violence is both predictable at home from someone known and an ever-present uncertainty from strangers in public; when the threat in the home remains even when the aggressor is out of doors, then women's fear of crime in general, and sexual assault in particular, is logical. The value of the 'wife rape' survey when analysed in the context of localized crime surveys which concentrate on 'outside crime' is to draw attention to the link between the public and private space dimensions of crime and illustrate how spatial structures and the social meanings attached to them have a central relevance in explaining the commission and impact of crime upon women. Households are physical structures which are also an extension of one's self, role, ego and social position in the world. Household space carries particular gender-based meaning and provides the physical and social context for understanding and interpreting the 'seriousness' of crime, hidden from public view, which occurs within it.

A recent case study of elderly people (the overwhelming majority of whom were women) in the inner city also illustrates the way in which variables of gender and age combine to alter the impact and evaluation of crime which occurs to physical structure of households. The elderly women in the survey were subjected to frequent acts of petty vandalism to their property. For example, plant pots and windows broken, garden plants damaged, men urinating in the gardens and porches, and so on. But the meaning attached to the household meant that vandalism and abusive behaviour was not regarded so much as a trivial property offence but as a serious domestic and personal offence because the acts themselves violated the privacy, control and security of home and community space. Moreover, it appeared that these 'minor' offences functioned as 'trigger' crimes, evoking fear of more serious crimes such as street robbery and assault. The frequency of the incidents also affected the interpretation put on them by respondents. As the report concludes:

> If crime is a rare occurrence it is possible to write it off as a piece of bad luck. But if crime and nuisance occur frequently then they are experienced as a grave injustice and unfairness. They symobolise a world out of control and are a powerful reminder of one's social and political impotence and physical vulnerability. Objectively they may be categorised as less serious crime but subjectively they are the way aggression is experienced in the world. ... Quite simply, people do not experience crime in terms of atomised, hierarchical, legal crime categories ... human evaluation of seriousness is related to the way crimes are grouped together and experienced in people's social psychology, physical, social and temporal space.
>
> (Painter 1989b: 108).

CONCLUSION

This chapter aimed to provide an empirical base to show the validity of women's fear of crime in urban areas. The risks are not simply subjective perceptions: they are real. Women's lives are controlled by the reality of male violence and crime against women in a patriarchal society is not atypical. The data also underscore the importance of relating the objective and subjective dimensions of human evaluation of antisocial and criminal incidents. In relation to crime there can be no 'objective' or 'actual' level of risk. Crime has a specifically subjective and evaluative element (Young 1988). Crime and fear of crime do not exist outside a social and spatial and temporal context; they depend upon definition whereby the act takes on a specific social meaning. And that meaning is governed by a number of dimensions which affect judgements as to 'seriousness'. The nature of the relationship between the male offender and the woman at the time of the incident, the nature of the incident itself, the physical setting of the home, street, work-place, the frequency of the incident, whether it is a once-only experience or part of a continuous cycle of abuse over time, and the perceived extent of the threat by the woman at the time and in retrospect, how it connected with other experiences, are all dimensions which affect definitions as to seriousness and impact.

How crime is defined and evaluated embodies a subjective assessment or judgement about which particular types of incidents cause greater physical harm or negative effects. Obviously rape is more serious in terms of physical harm done than a verbal obscenity hurled across the street. But as Kelly (1987) points out, rape and domestic violence actually affect a minority of women (so in terms of frequency and extent could be regarded as less serious) whereas the constant threat of sexual violence in public space which every woman experiences could be regarded as more serious because of the tangible effects it has on restricting women's access to public areas, day and night, throughout their lifetime.

Empirical information is crucial in debunking the conventional wisdom that women are not objectively at risk. But it is also evident that women's experiences and fear of crime in time and space is a complex matter and cannot simply be read off a computer printout organized around discrete legal, hierarchical categories differentiating serious and non-serious crime over a twelve-month period. Underlying all women's experiences of crime and the threat of crime is a common social process: that of social control through male domination. It is a form of control which operates across all spatial contexts which interconnect and in turn, reinforce existing social divisions. Crime against women is a social relationship which cannot be theorized autonomously from its spatial and temporal context and though the theoretical strands to this chapter are, admittedly, undeveloped the empirical evidence does point to connections between victimization, patriarchal and class relationships and the usage, control, form and values attached to physical and spatial aspects of the urban environment and crime which occurs within it. It also illustrates the immense variation in definitions, risks and fears of crime between individuals and within communities over small geographical areas. So much so, that it is reasonable to conclude that men and women live in worlds which experientially, even in the confined space of the inner city, are aeons apart.

Acknowledgements

I am extremely grateful to Chris Nuttall, Sandra Walklate, the 'Two Davids' (Evans and Herbert) and Jock Young for their comments on an early draft of this chapter.

NOTES

1 The boroughs and estates referred to in the text are not necessarily the same types of areas. The term 'borough' refers to an administrative and political boundary which is still retained in London. The borough of Hammersmith and Fulham and the borough of Islington are both inner city areas with Labour councils and predominantly Labour Members of Parliament (MPs). Fulham has a Conservative MP. The areas are typical of mixed inner metropolitan areas classified by ACORN Neighbour-hood Group 4, containing multi-occupied older housing; cosmopolitan owner-occupied terraces; multi-let housing; better off/gentrified areas. The estates referred to differ in type. For example, the West Kensington estate built in the 1960s would be best classified as one of the better estates in Fulham and would aproach ACORN Neighbourhood Group E, Category 1. By contrast, the Ladywood estate, located in Birmingham, would be best classified as an extremely deprived, run down and impoverished estate, i.e. ACORN Neighbourhood Group G, Category 3.

2 The term 'harassment' appears frequently throughout the text and so it may be helpful to clarify what is meant by this. Distinctions are drawn between general harassment and sexual harassment, though it has to be admitted that these incidents may overlap both in substance and indivi-dual interpretation put upon them. Thus, whereas men and women may be shouted at, verbally abused, stared at, and so on, when the abuse is directed toward a woman it is frequently directed, at her sexuality. In the surveys women were asked if they had been 'sexually harassed' and it was left to women respondents to interpret what was meant by this. The question on annoying behaviour or general harassment was asked of men and women and was broken down into the categories of behaviour outlined in Table 9.3.

REFERENCES

Baldwin, J. and Bottoms, A. (1976) *The Urban Criminal*, Tavistock: London.

Bottoms, A. and Wiles, P. (1988) 'Crime and housing policy: a framework for crime prevention analysis', in T. Hope and M. Shaw (eds) *Communities and Crime Reduction*, London: HMSO.

Crawford, A., Jones, T., Woodhouse, T., Young, J. (1990) *The Second Islington Crime Survey*, Centre for Criminology, Middlesex Polytechnic.

Forrester, D., Chatterton, M., Pease, K. (1988) *The Kirkholt Burglary Prevention Project, Rochdale*, Crime Prevention Unit Paper 13, London: Home Office.

Giddens, A. (1985) 'Time, space and regionalisation', in D. Gregory and J. Urry (eds) *Social Relations and Spatial Structures*, London: Macmillan.

Gottfredson, M. (1984) *Victims of Crime: The Dimensions of Risk*, Home Office Research Paper 81, London: Home Office.

Green, E., Hebron, S. and Woodward, D. (1987) 'Women, leisure and social controls, in J. Hanmer and M. Maynard (eds) *Women, Violence and Social Control*, London: Macmillan.

Gregory, D. and Urry, J. (1985) *Social Relations and Spatial Structures*, London: Macmillan.

Hanmer, J. and Saunders, S. (1984) *Well-Founded Fear*, London: Hutchinson.

Hope, T. and Hough, M. (1988) 'Area crime and incivilities: a profile from the British crime survey', in T. Hope and M. Shaw (eds) *Communities and Crime Reduction*, London: HMSO.

Hough, M. and Mayhew, P. (1983) *The British Crime Survey*, Home Office Research Study 76, London: Home Office.

—— (1985) *Taking Account of Crime*, Home Office Research Study 85, London: Home Office.

Jacobs, J. (1961) *Death and Life of Great American Cities*, New York: Random House.

Jones, T., Maclean, B., Young, J. (1986) *The Islington Crime Survey*, Aldershot: Gower.

Kelly, L. (1987) 'The continuum of sexual violence', in J. Hanmer and M. Maynard (eds) *Women, Violence and Social Control*, London: Macmillan.

Lea, J. and Young, J. (1984) *What is to be Done about Law and Order?*, Harmondsworth: Penguin.

Lea, J., Jones, T., Woodhouse, T. and Young, J. (1988) *Preventing Crime: The Hilldrop Environmental Improvement Survey*, First Report, Centre for Criminology, Middlesex Polytechnic.

Lewis, D. and Maxfield, M. (1980) 'Fear in the neighbourhoods: an investigation of the impact of crime', *Journal of Research in Crime and Delinquency* 17: 140–59.

Lewis, D. and Salem, G. (1981) 'Community crime prevention: an analysis of a developing strategy', *Crime and Delinquency* 27: 405–21.

McNeill, S. (1987) 'Flashing: its effects on women', in J. Hanmer and M. Maynard (eds) *Women, Violence and Social Control*, London: Macmillan.

Maguire, M. and Bennett, T. (1982) *Burglary in a Dwelling*, London: Heinemann.

Mawby, R. (1986) 'Contrasting measurements of area crime rates', in R. Miyazawak and M. Ohya (eds) *Victimology in Comparative Perspective*, Tokyo: Seibundo.

Mayhew, H. (1861) 'A visit to the rookery of St. Giles and its neighbourhood', reprinted in M. Fitzgerald *et al.* (1981) *Crime and Society*, London: Routledge & Kegan Paul.

Mayhew, P., Clarke, R.V.G., Burrows, J.N., Hough, J.M. and Winchester, S.W.C. (1979) *Crime in Public View*, Home Office Research Study 49, London: HMSO.

Mayhew, P., Elliott, D. and Dowds, L. (1989) *The 1988 British Crime Survey*, Home Office Research Study 111, London: HMSO.

Morris, T. (1957) *The Criminal Area: A Study in Social Ecology*, London: Routledge & Kegan Paul.

Newman, O. (1972) *Defensible Space*, New York: Macmillan.

Painter, K. (1988) *Lighting and Crime Prevention: The Edmonton Project*, Centre for Criminology, Middlesex Polytechnic.

—— (1989a) *Lighting and Crime Prevention for Community Safety: The Tower Hamlets Study*, Centre for Criminology, Middlesex Polytechnic.

—— (1989b) *Crime Prevention and Public Lighting with Special Focus on Elderly People*, Centre for Criminology, Middlesex Polytechnic.

—— (1990) *Wife Rape, Marriage and the Law*, Manchester: Manchester University Press.

Painter, K., Lea, J., Woodhouse, T. and Young, J. (1989a) *The Hammersmith and Fulham Crime and Policing Survey*, Centre for Criminology, Middlesex Polytechnic.

—— (1989b) *The West Kensington Estate Survey*, Centre for Criminology, Middlesex Polytechnic.

Painter, K., Woodhouse, T. and Young, J. (1990) *The Ladywood Crime and Community Safety Survey*, Centre for Criminology, Middlesex Polytechnic.

Park, R.E., Burgess, E.W. and McKenzie, R.D. (1925) *The City*, Chicago: University of Chicago Press.

Quetelet, A. (1869) 'Physique Sociale', quoted in L. Radzinowicz (1966) *Ideology and Crime: A Study of Crime in its Social and Historical Context*, London: Heinemann.

Shaw, C. and McKay, H. (1942) *Juvenile Delinquency in Urban Areas*, Chicago: University of Chicago Press.

Skogan, W. (1987) *Disorder and Community Decline*, Centre for Urban Affairs and Policy Research, Northwestern University, Evanston, Ill.

Smith, S. (1989a) 'Social relations, neighbourhood structure and the fear of crime in Britain', in D. Evans and D. Herbert (eds) *The Geography of Crime*, London: Routledge.

—— (1989b) 'The challenge of urban crime', in D. Herbert and D. Smith (eds) *Social Problems and the City*, Oxford: Oxford University Press.

Stanko, E. (1985) *Intimate Intrusions*, London: Routledge & Kegan Paul.

—— (1987) 'Typical violence, normal precaution: men, women and interpersonal violence', in J. Hanmer and M. Maynard (eds) *Women, Violence and Social Control*, London: Macmillan.

—— (1988) 'Hidden violence against women', in M. Maguire and J. Pointing (eds) *Victims of Crime: A New Deal?*, Milton Keynes: Open University Press.

Taylor, D.G., Taub, R.P. and Peterson, B.L. (1986) 'Crime, community organisation and causes of community decline', in R.M. Figlio, S. Hakim and G.F. Rengert (eds) *Metropolitan Crime Patterns*, New York: Criminal Justice Press.

Walsh, D. (1983) *Break-ins: Burglary from Private Houses*, London: Constable.

Warr, M. (1985) 'Fear of rape among urban women', *Social Problems*, 32.

Young, J. (1988) 'Risk of crime and fear of crime: a realist critique of survey-based assumptions', in M. Maguire and J. Pointing (eds) *Victims of Crime: A New Deal?*, Milton Keynes: Open University Press.

10

A PLACE FOR EVERY CRIME AND EVERY CRIME IN ITS PLACE

An alternative perspective on crime displacement

Robert Barr and Ken Pease

The intention of this chapter is to substitute one way of thinking about crime movements for another. While it includes an analysis of victimization data from the British Crime Survey, its principal aim is to outline a novel approach to the analysis of crime data and to establish an agenda for research and its application, which is set out at the end of the chapter. The need for presenting the issue in general terms first is because of the central importance of how we characterize crime movements for our assessment of crime control policy in general and policing in particular. A shift of perspective is a necessary condition for the development of applied research in the area.

CRIME MOVEMENTS AND CRIME STATISTICS

The recording of criminal incidents takes up a substantial propor-tion of police time. There is a slight, albeit increasing, tendency to use such data immediately for operational purposes. However, its more routine use is after aggregation, for incorporation into statis-tical reports for senior officers, the press and the Home Office. Such aggregations are presented as a social indicator or moral barometer, and are recognized as having political implications for local and national governments. Ratios between aggregated figures (notably the clearance rate, or ratio of crimes cleared to crimes known) give service as measures of police performance. Despite all the acknow-ledged shortcomings of such figures (see Bottomley and Pease 1986 for an exposition of these), social scientists continue to seek to explain the revealed patterns of criminal activity by relating aggregated crime data to external variables such as the demographic characteristics of the local populations set out in the census, or to economic indicators such as unemployment figures. For many purposes, officially recorded data do seem to provide a reasonable proxy of crime suffered, as long as they are appropriately grossed up according to the findings of victimization surveys. This process requires victimization surveys of a high level of sophistication (see Skogan 1990) but is possible. The day when it was possible to dismiss statistics of crimes known as totally meaningless have thankfully gone (except among the ideologically committed on the Left).

The important problem with criminal statistics, which is implicit in the foregoing, is not any inherent and general meaninglessness, but their lack of articulation with the immediate concerns of justice personnel. This is primarily because they are based on discrete periods of the past, and emerging patterns are usually not sought at all or are available too late for effective action. There is an element of bolting the door after the stable has been burgled! This is not to say that useful patterns are never detected, but that this tends to be the result of perceptive individual officers rather than more formal analysis of patterns of recorded crime. A further difficulty is that the units across which statistics are aggregated frequently relate poorly to the pattern of crimes when plotted individually. Conventional statistics are subject to the ecological fallacy because the characteristics of individual offenders or victims are inferred, possibly erroneously, from the aggregate characteristics of the population from which they are drawn.

One crucial instance of the lack of articulation between police concerns and routine statistical summaries concerns the distribution of crimes. This is because, acting as an albatross around the neck of purposive crime prevention, has been the 'displacement hypothesis'. This hypothesis suggests, in its most extreme form, that whenever a criminal opportunity is blocked, the would-be criminal does something else. Such a view is favoured by those who see criminal behaviour as a manifestation of underlying social or psychological disorder. Four types of displacement responses have been identified (Hakim and Rengert 1981):

1 *temporal* – committing the intended crime at a different time
2 *spatial* – committing the intended crime in a different place

3 *tactical* – committing the same crime in a different way
4 *target* – committing a different crime from that originally intended.

The principle underlying each of the above forms of displacement is that the incentive to commit a crime is strong enough to encourage a change of strategy if the primary intention is thwarted. Conventional crime statistics are of little use in measuring whether displacement takes place. Temporal displacements will usually not find their way from incident reports to statistics aggregated across long time periods. Spatial displacements will seldom interact with the boundaries of statistical areas in such a way as to be recognizable. Tactical and crime type displacements also require sensitive temporal monitoring. The result of all this has been that displacement has been almost exclusively studied in the context of change programmes, and especially crime prevention programmes. Displacement has not therefore been seen as a constant flow of crime types and circumstances, in the way developed later in this chapter.

Paradoxically, the behaviour most intensively analysed by criminologists in relation to the displacement hypothesis is not a crime: it is suicide. The shift from (toxic) manufactured to (non-toxic) natural gas in the United Kingdom led to a reduction in the rate of suicide (Clarke and Mayhew 1987). The argument by analogy is that if a fatal decision like the taking of one's own life can be reversed by changing opportunities, then *a fortiori* so will the less momentous decision to commit crime. Thus crime displacement will not be total when an opportunity is blocked, and preventive action may well yield reductions in crime. Similarly, the introduction of motor cycle helmets reduces the number of motor-cycle theft, but these crimes seem not to be displaced to car thefts (Mayhew *et al.* forthcoming). More typically (although less intensively), displacement has been investigated as part of crime prevention programmes. For example, the introduction of steering column locks for new cars in the UK and its effects on the taking of older vehicles provides an early example in the literature (Mayhew *et al.* 1986). The modestly substantial British research of this kind is reviewed in Barr and Pease (1990), and there is North American research of the same kind (Chaiken *et al.* 1974; Stenzel 1977; McIver 1980). This is not the appropriate place in which to re-review the work. All we need to note is that displacement of varying extents has been demonstrated, but that the claim that displacement is total is never, and never can be, precluded by research. This is, *inter alia*, because routine data on crime are always time and place limited. Thus there is always a time or place (or crime type) to which a crime may have been displaced which lies outside the scope of crimes measured. Reducing street robbery may increase shop theft, burglary or fraud. If the primary gratification is violence, it may transfer to spouse abuse. If a small part of prevented crime goes into other kinds of crime, the total merges into measurement error. Researchers are reduced to arguments like someone prevented from stealing a motor bike to get home late at night might mug someone for the taxi-fare instead or might hire a taxi and run off without paying. For most offenders, however, the choice-structuring properties of these offences may be too dissimilar from motor-cycle theft to provide realistic alternatives. For instance, they involve direct confrontation with the victim and a greater risk of apprehension. Mugging also entails a willingness to engage in violence and correspondingly more severe penalties. Given that displacement to offences of these kinds is likely to be minimal, it seems reasonable to confine the search for displacement to thefts of cars and bicycles (Mayhew *et al.* forthcoming). Common-sense restrictions of this kind have to be made to render the conventional study of displacement possible. Equally inevitable is the rejoinder that displacement outside the selected alternatives could have occurred.

It is important to stress that we are not pessimists about the effects of crime prevention programmes. We do not believe that all crime 'prevented' is merely moved somewhere else to happen. We have elsewhere preferred the usage crime deflection to crime displacement because it focuses on the achievement of moving a crime from a chosen target, whereas displacement focuses on its movement to somewhere, something or somewhen else. That was an attempt to escape from the paralysing influence of the displacement hypothesis and it remains the usage we prefer, although we have avoided it in this chapter because the topic is more familiar when sailing under the flag of displacement. However, when all the optimistic changes of vocabulary are done, the fact remains that no one can demonstrate that displacement/deflection is less than total, and we are in consequence profoundly dissatisfied with the conventional approach to displacement issues. We link the dissatisfaction to the typical context of displacement measurement, namely the measurement of the effects of change programmes, rather than the measurement of crime distributions over the relevant dimensions.

FROM DISPLACEMENT TO PLACEMENT

The concept of displacement provides an interesting and important perspective on crime and policing. It is a perspective with an inherently spatial component and provides an opportunity to apply a range of analytical techniques developed elsewhere in the spatial disciplines. Traditional crime statistics encourage statistical mapping, the establishment of ecological relationships and a search for highly criminalized 'hot spots' where police activity may need to be concentrated. Displacement hypotheses suggest the application of spatial diffusion theories (more frequently employed to analyse the spread of infectious diseases), temporal studies that recognize the functional differences in the same places at different times, and behavioural approaches to both victim and offender conduct. While intellectually challenging, and of some academic interest, the pursuit of the displacement hypothesis is unlikely to have a direct impact on policing practice or even the collection of police statistics. So is there an approach that can be derived from the displacement concept which may be of more practical utility?

Displacement is usually taken to involve an existing intention in an individual or group of potential offenders to commit a particular offence in a particular place at a particular time. Two different crime deflection or contagion effects may be introduced to expand the discussion towards our goal.

First, an additional displacement effect has been identified (Barr and Pease 1990) which extends the concept. This can be termed 'perpetrator displacement'. An extreme example is that of the drug courier. Arresting and imprisoning a single courier has little effect on the traffic because the trafficker can offer such high rewards to couriers that there is an inexhaustible supply of volunteers. The value added to the product by the courier is so high that this may lead the trafficker to employ larger numbers of couriers to increase the probability of a shipment getting through rather than to concentrate on any individual shipment. Less dramatic examples may include the impact of a police car apprehending a speeding motorist on a motorway. The sight of the parked police car may encourage other motorists to speed on the assumption that their risk of apprehension is reduced while the stationary patrol car is otherwise occupied.

Second, a fission effect can be recognized. This can include copycat offences such as the spate of social worker impersonations, graffiti artists or even inner city rioters. It may include revenge acts which are themselves offences, such as tit-for-tat and punishment shootings. The prevention of an offender re-offending (for a while at least) in itself triggers a series of protest offences by alternative perpetrators (public disorder to 'free the Hicksville n') or the same perpetrators elsewhere (for example the Strangeways riot).

An important change of perspective is brought about by the extension of the simple concept of one-for-one crime displacement to one where the deflection of crimes can trigger a larger or smaller number of similar or different crimes by the same or different people. This process we shall term 'crime flux'. It is clear that our earlier preference for the term 'deflection' did not go far enough, since (by analogy) a blow deflected is still just one blow. Crime flux does not carry the same restrictive sense. The analysis of crime flux puts the spotlight on crime movements across offenders and crime types, as well as within the more limited range of the conventional approach. However, we shall return to the customary usage (crime displacement) but ask that it be understood to refer to the liberalized form developed above, where changes across offenders, general crime types and changes in numbers of crimes also find a place.

If displacement (crime flux) takes place, its consequences may be classified as benign, malign or neutral. Benign displacement involves a less serious offence (or offences) being committed instead of a more serious one, an act of similar seriousness being moved to a victim or victims for whom the act will have less damaging consequences, or even a non-criminal act instead of a criminal one (more usually referred to as desistence). Malign displacement involves a shift to more serious offence(s), or to offences directed where they will have worse consequences. To assess the effect of potential displacements it is necessary to classify potential displacement effects. This can be done on an individual crime basis, dealing with each alternative crime and its undesirability. Alternatively it can be done on a holistic basis, where a particular profile of crime is considered to be more or less undesirable than the present pattern. Clearly the latter is both more realistic, and truer to the nature of the concept of crime flux.

Presuming for the moment that we can measure crime flux adequately, the change of emphasis raises the following question that is seldom if ever asked, at least officially, 'if crime is not preventable, where, to whom and by what means should it take place?' This is the placement rather than the displacement question. Once emphasis shifts to crime placement we cease to be obsessively concerned about the absolute quantity of crime and its distribution, the present focus of crime analysis. While having the same objective, which is to prevent harms through crime, a different approach would be taken to crime control policy. New questions are raised. Geographically we might ask 'Why has this crime occurred here rather than somewhere else?', 'By taking defensive action where might we move this crime to?' 'Would we prefer it there?' 'What kind of alternative perpetrators and crime opportunities are available?' The issue presented as one of apportionment changes the emphasis from existing distributions to the generation of normative distributions against which we can judge future changes in crime pattern. The value of such an approach is that benign displacement becomes a valuable adjunct to crime prevention rather than its adversary. Because our approach when presented verbally has been often misunderstood, it is important to restate that the preferred aim is crime reduction. The difference is that the partial, or if benign, the total displacement of crime would itself be a desirable objective. How then might we arrive at our normative pattern of 'acceptable' crime, the preferable purgatory which is a staging post between the present crime pattern to the questionable paradise of a society with no crime, and perhaps the closest we can get to that paradise. There are three elements which already feature as part of the judicial determination of appropriate sentence. These are the vulnerability of the victim, the consequences for the victim and the fear of becoming a victim in the rest of society. The first two at least could form the basis of the analysis of crime flux. The reason for excluding the third, crime fear, will be set out below.

At the individual level it is possible, albeit imperfectly, to measure or infer the potential vulnerability of the individual to suffering a particular crime. Major victimization surveys such as the British Crime Survey (BCS), as well as aggregate crime statistics, make it possible to calculate probability of criminal victimization for any individual, based on their age, sex, place of residence and ethnic group. This has been done extensively in relation to the BCS by Gottfredson (1986), and by a variety of North American scholars.

It is more difficult, but possible, to estimate the potential consequences of crime for an individual, and much of the literature of victimology addresses the question of consequences. However, the ways in which social variables interact to yield consequences makes generalization difficult. For example if a well-paid suburban commuter has her car stolen she is likely to suffer a degree of inconvenience and irritation and a limited degree of financial penalty. By contrast if a poorly skilled worker on an outlying housing estate has his car stolen, on which he relies for travel to work, the consequences may be the loss of a job.

The fear of crime, rather than crime itself, has been regarded as an important third factor in recent literature. It is the fear of crime that is exploited, or to put it more fairly, responded to by politicians in the law and order debate. In turn an important function of the police and the judiciary, it could be contended, is to reduce not crime itself, but the public's fear of crime victimization. The fear of crime can be used in crime prevention programmes as a means of encouraging defensive measures. It is frequently used by politicians as a critical tool againt opponents, although the kinds of fear which come to the fore are interestingly different as between the major parties. The difficulty that this presents to the policy-makers is that a pressure exists to respond to the fear pattern rather than the actual distribution of crimes, or of vulnerable individuals. The emphasis on fear of crime is criticized particularly by the left realist school of criminology (see Jones *et al*. 1986). The best established generalization about crime fear is that the gender/age group which is most afraid of falling victim is least likely to suffer a crime. Thus, implicitly, the fear is taken to be irrational. Perhaps the second best established relevant generalization is that levels of expected victimization when averaged across a western country are very low. Thus, implicitly, crime is not a problem to fear greatly. In essence, the left realists' argument concentrates on the second point, which they contend is spurious. Their local victim surveys suggest that many crimes are not captured by the national surveys, and that in some areas the fearful groups have every reason to be so. This is not the place to expatiate upon the methodological considerations which underpin the conflict. Rather, it is possible to summarize the alternatives and take a position. The position taken is that crime fear can be regarded as an arithmetic product of (perceived) probability of victimization and severity of the consequences of victimization. Thus crime fear would prove manipulable by changes in either of the two contributory variables. In practice, it would be relieved by reducing the probability of victimization or by improving victim support services (or both). Attempts to manipulate fear by other means would be doomed to failure. There are a number of features of this position that we find attractive.

1 It avoids the attempts to pathologize fear in advance of the fear being demonstrated to be neurotic in origin.
2 It gives proper emphasis to crime consequences, which may be the key to understanding crime fear. An elderly woman may have a low probability of being assaulted or robbed, but if the consequences were to be disabling, the fear would be rational.
3 It avoids the mounting of mere publicity campaigns to reduce crime fear.
4 The consequences of being wrong are not severe. If the position is wrong, the worst that happens is that efforts are inappropriately concentrated on reducing crimes or crime consequences. The position set out is a strong one. In a weaker variant, it is difficult to imagine that crime fear is totally unaffected by changes in crime or crime consequences, so addressing these issues would be relevant.

In summary, fear is to be excluded from the analysis of crime flux because assertions of its existence independent of relevant crime variables prematurely demeans and distorts public sentiments.

THE SPATIAL DIMENSION

If the management of crime distribution is to take account of the factors we have discussed, we need methods of establishing the expected crime norms for operational areas and for the monitoring of changes in the pattern of individual types of crimes over time. At first sight there appears to be a tension between an interest in the individual victim or offender and his or her personal characteristics that is characteristic of the criminologist and the aggregate, area-based approach usually adopted by the statistician or geographer. It is important to recognize the extent to which the demographic or other statistical characteristics of spatial units are used as ecological surrogates for the characteristics of individuals living in them. The problem operates in two ways. The classical ecological fallacy is based on taking a relationship that occurs at the aggregation unit level and applying it to individuals within that unit. For example there may be a negative correlation between car ownership and the number of reported car thefts at the level of the police beat or division. It should not, necessarily, be inferred from such an association that individuals not owning a car are likely to steal one from their own neighbourhood! However, there are more fruitful methods available for inferring the characteristics of areas from the characteristics of individuals within them. The technique of micro-simulation (Clarke and Wilson 1987; Clarke and Holm 1987) has been used successfully to model the demand for health services or shopping behaviour of areas by applying a simulation model to the demographic characteristics of the population of an area which is driven by the probabilities of particular events. Such an approach could be used to combine data on victimization from police and survey records with the underlying characteristics of the population to drive a simulation model which yields expected victimization rates, for a set of areas, for a range of crimes. Not only can such a model generate 'crime norms' against which actual patterns may be judged but also it could have predictive value. For example it could take account of known factors such as multiple victimization or other known social or economic correlates of crime incidence and incorporate these to predict, more accurately, the expected levels of crime in particular areas. The importance of such a model, or set of models, would be to provide a mechanism for the incorporation of well-founded criminological theory into an operational management tool for the police. The acceptance of such models, however elegant, is likely to be problematic because the concept of a 'norm' is inevitably based on a value stance that leads to a preferred pattern of crimes. It involves answering our hard questions. The traditional notion of displacement, or deflection as we have termed it, involves the substitution of a known crime on a known target to an unspecified crime against an unspecified target – or with luck no crime at all. The concept of modelled 'norms', while not specifying the alternative targets absolutely, does bring into focus the possible, or the expected, consequences of particular crime prevention strategies.

The discussion so far has focused on a novel spatial approach which allows us to take advantage of the wealth of individual incident data available from the police and of victimization data from crime surveys. A complementary approach to micro-simulation is to see the extent of police expertise in crime flux. To our knowledge, no one has asked police-officers to predict the effect of changes on crime flux. (Asked after the event, many police-officers seem totally confident that they could have predicted how crimes were displaced. This is curiously often outside the areas, times and types of crime which the researcher has chosen to measure.) It is possible, and would be exciting, if police-officers could be recruited as domain experts for the development and testing of the microsimulation approach described above.

As we have outlined above, traditional aggregate approaches have been concerned with the absolute quantity, and the distribution over space of individual crimes. We have proposed (Barr and Pease 1990) a conceptual model (Figure 10.1) that allows us to classify individual crimes and to discuss desirable and undesirable changes in their distribution. This model is presented as a triangle. Introducing a time dimension would yield a pyramid. Positions on the triangle/pyramid represent distributions of crime. In the spirit of the above discussion, we would intend that in any operational use, the relevant variable for the model would be the product of a standardized measure of crime harm, rather than a crime count. However, for purposes of exposition, we restrict ourselves to a crime count here. The pinnacle of the triangle represents the notional, if unattainable, aim of eliminating crimes completely. Any

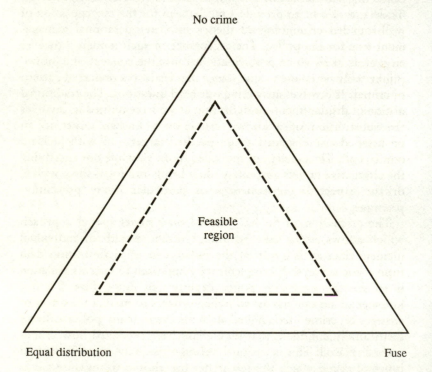

Figure 10.1 Conceptual model of crime distribution

individual type of crime is located in this space on the basis of two variables, its prevalence which locates it in the vertical axis and its concentration in space along the bottom axis. Its concentration across time would be the extra dimension in a pyramidal version. The model as presented would be as devised by a geographer. The criminologists may be equally interested in concentration among particular demographic, racial or social groups. A helpful compromise would be to use a geo-demographic classification such as ACORN, A Classification of Residential Neighbourhoods (CACI 1988), which can act both as a proxy for different types of residential environment, usually separated in space, and different social groups. The basic premise of the model is that the horizontal shift of an individual crime type may be as significant as a vertical shift. We have termed the extreme of concentration the 'fuse'. This is intended as analogous to the electrical circuit, in which a deliberate weak point is introduced, so that when a power surge burns out part of the circuit, the search for its location will be brief. There are clearly cases, admitted or otherwise, in which policing problems are reduced by having fuse areas or groups. Prostitution is an obvious example. A fuse area allows oversight and regulation of an illegal trade which is not likely to be eliminated totally. When public outcry becomes too great, a focused operation can be mounted. Women know which parts of a town or city are to be avoided, although they will ask why they should have to avoid anywhere. The point here is simply that there are some advantages to fuses rather than uniform distribution of a crime problem. Similarly, 'fuse' unruly pubs are useful in allowing the quiet drinker to avoid assault by going elsewhere, rather than having unruly drinkers spread like an unwholesome yeast across an area's licensed premises. Similar arguments may be made in relation to the drug trade.

To operationalize the conceptual model we can use two parameters. The first is a prevalence measure for any individual crime type – defined as the actual number of victims divided by the total number of potential victims. It will thus vary between 0 and 1. The measure of concentration that can be used is the Gini coefficient which also varies from 0 to 1, 0 denoting a completely equal distribution and 1 a total concentration of the crime in one area or on one type of victim. Figure 10.2 divides this operationalized graph into four zones (one of which, as explained below, would never occur in practice under the simple model presented here). It is

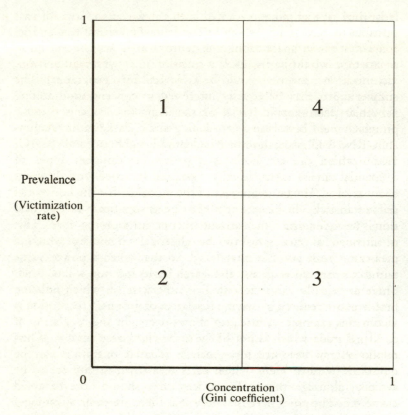

Figure 10.2   A classification of crime distribution

envisaged (but is not necessary) that certain crime types will usually fall into one zone, and the figure is illustrated on that assumption. Predominantly Type 1 crimes would be characterized by a high prevalence and a relatively equal distribution (low concentration), Type 2 crimes by low prevalence and low concentration, and Type 3 crimes by low prevalence and high concentration. When actual crimes are plotted Type 4 (high concentration and high prevalence) could not occur in this model because statistically it is not possible to have high prevalence, where many of the potential victims suffer a crime, and high concentration. This is because the pool of victims necessary to create a high prevalence score is not available in any one area or group of people. This would not be true if one took number of victimizations per victim into account. In other words, if frequency of victimization were taken into account, it would be possible to have high prevalence and high concentration, if the frequency of victimization among victims were markedly unequal – as indeed it is (see Sparks 1981 for a classic statement about multiple victimization). However, we shall restrict ourselves here to the simpler approach, in which only three crime types are possible. This is both for ease of exposition and because the British Crime Survey, on which most British analysts would wish to explore the relationships described, poses severe restrictions on the analysis of multiple victimization.

Crimes can be classified into the three practical types by the analysis of victimization surveys such as the BCS or from routine crime statistics. The main problem in operationalizing the classification is to obtain data which are sufficiently well referenced spatially, or across social groups, to allow the calculation of the concentration measure. Crime statistics at divisional level, for example, will hide within them many local patterns of high and low concentration. In order to be able to operationalize such classifications effectively, better geo-referencing systems and better systems of recording victim characteristics are required.

A preliminary analysis of data from the BCS shows a strong relationship between crime prevalence and concentration. Screening questions on respondent's experience, as victims, of criminal offences provide data on some fifteen types of successful or attempted crime. The responses, weighted to balance the sample between inner city, urban and rural areas can be used directly to calculate a prevalence score for each crime type. Spatial, or social, concentration can be calculated by examining the differing prevalance rates across a set of areas, or social groups. The sample for the BCS is well stratified and based on parliamentary constituencies, wards and polling districts. Spatial concentration could be calculated on the basis of the sampling areas, however, in 1984, the CACI ACORN code was included for the home address of each respondent. We chose to use this as a basis for our analysis of concentration: each crime was analysed across eleven ACORN neighbourhood groups, characterized by descriptions such as 'Affluent Suburban Housing' or 'Mixed Inner Metropolitan Areas'. ACORN groups have the advantages of being reasonably spatially coherent and of reflecting the underlying social environment. Concentration was measured using the Gini coefficient.

The distribution of crimes, when plotted on the operational axes

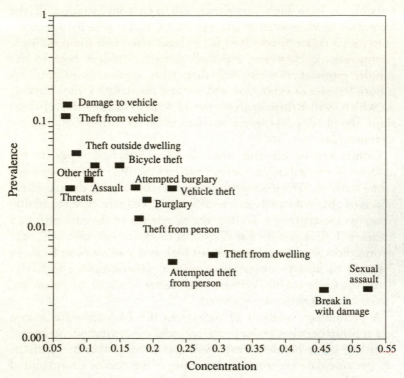

Figure 10.3 Relationship between the prevalence and concentration of
selected crimes
Source: NOP (1984) *British Crime Survey* Victim Data

we have proposed, shows a strongly 'L' shaped distribution. (Figure 10.3; Table 10.1). For plotting and regression the distribution can be straightened by taking the Log of prevalence, or, the Log of both prevalence and concentration. A Log Log regression yields a correlation of −0.9 (Table 10.2). Showing a strong relationship between prevalence and concentration. While this relationship may appear self evident, because, at the extremes when prevalence approaches 1 concentration must approach 0 and as prevalence approaches 0 concentration must approach 1, it should be noted that such extremes of prevalence are never approached for the commoner crimes which we are concerned with. A best fit curve through out set of points will pass through our Type 1, 2 and 3 zones. We are concerned with the position of individual crimes

Table 10.1 Victimization rates and concentrations

Crime	Prevalence	Concentration
Vehicle theft	0.024	0.232
Damage to vehicle	0.146	0.073
Theft from vehicle	0.114	0.069
Bicycle theft	0.040	0.151
Burglary	0.019	0.193
Break in with damage	0.003	0.457
Attempted burglary	0.025	0.175
Theft from dwelling	0.006	0.292
Theft outside dwelling	0.052	0.085
Theft from person	0.013	0.180
Attempted theft	0.005	0.231
Other theft	0.040	0.114
Assault	0.029	0.104
Threats	0.024	0.077
Sexual assault	0.003	0.525

Source: NOP (1984) British Crime Survey

$$\text{Prevalence} = \frac{\text{Respondents reporting one or more victim episodes}}{\text{Total number of Respondents to question}}$$

Concentration = Gini coefficient

Table 10.2 Regression of Log concentration (dependent variable) on Log prevalence (independent variable)

Regression output	
Constant	−1.60816
Std Err of Y Est	0.122626
R	−0.905635
R Squared	0.820174
No. of Observations	15
Degrees of Freedom	13
X Coefficient(s)	−0.48082954
Std Err of Coef.	0.062444269

along this curve and their divergence from it. These characteristics of the individual crime and its change of pattern over time provide the basis for our concept of crime flux (discussed more fully below).

An inspection of the crime plot derived from the 1984 British Crime Survey shows that Type 1 crimes include crimes involving motor vehicles and other moveable property. Type 3 crimes include sexual assault and breaking in and causing damage. While both relatively rare both these crimes show an alarming concentration in

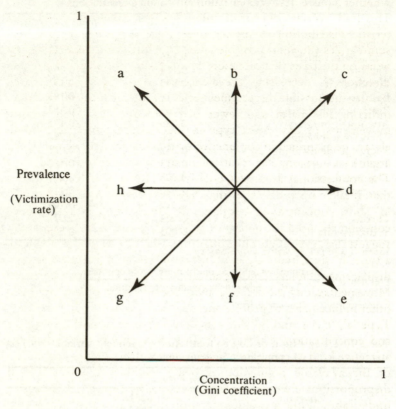

Figure 10.4 Crime flux

central urban areas, both poor and better off. Type 2 crimes, with low prevalence and low concentration, would include crimes such as theft (and attempted theft) from individuals and theft from dwellings. Type 1 crimes are characterized by damage to and theft from vehicles. It should be noted that the BCS records victims' experiences by the address of the victim, not by the location where the crime took place. A similar analysis of police incident records will yield rather different results. However the aim of this pilot exercise is to show that crimes can be classified usefully be relating prevalence to distribution.

The classification of crimes in this way allows a more rational approach to the distribution question. Rather than concentrating only on prevalence, it becomes possible to take decisions as to whether a more desirable distribution of the same crimes is feasible. This leads to our concept of crime flux, the change in the characteristic distribution of crimes over place and time, Figure 10.4 summarizes the eight possible changes. Most crime reports simply work on the basis that direction 'b' increasing prevalence is bad and direction 'f' decreasing prevalence is good. Such an approach ignores the possible displacement effect of the substitution of crimes inflicting less harm and fewer bad consequences. Thus vertical movements of any crime type in the classification space should always be examined in conjunction with horizontal movements. We argue that horizontal shifts are of equal importance to vertical ones. The conventional argument criticizing 'Home Watch' schemes is that, if taken up widely they may have an effect of producing a type 'd' shift, concentrating burglaries in areas which lack the social cohesion required for defensive action. Arguments of social equity suggest that, even if absolute quantities of crime cannot be reduced, a type 'h' shift, reducing concentration would constitute a benign displacement unless there is an argued case for deliberate fuses. Movements along the d–h line invite consideration of insurance and other inducements to protect oneself, and its distribution in society. Type 'a', 'c', 'e' and 'g' shifts include a change in both prevalence and concentration. The type 'c' shift may appear to be the least attractive kind of crime flux, since it means that a higher proportion of the available population is being victimized, and this falls disproportionately on some areas. Such a change may not always be undesirable. While a swamp operation leading to urban riots would be characterized by a type 'c' shift and may not be seen as an effective method of reducing urban crime, an effective drugs control policy may lead to an apparent increase in prevalence and concentration of the problem. In this case the apparent increase in prevalence is likely to be the result of revealing previously undetected offences, which can be controlled more easily if spatially concentrated, hence the experiments in certain European cities of allowing drug use and distribution in certain defined and monitored areas.

A pessimistic(?) view of 'Home Watch' may be that it achieves a type 'e' shift which reduces crime overall but concentrates it on the most vulnerable victims. Type 'g' shifts would usually, but not invariably, be seen as the most desirable, reducing and dispersing crime. If a policy of prioritizing work to match goals specified in terms of Figure 10.3 were to be adopted, the amount of effort expended to achieve further type 'g' shifts may often be seen not to be justified and other priorities should be pursued. For instance the scale of (differential) effort required to reduce both prevalence and concentration of income tax fraud may well be deemed disproportionate, as may the effort required to have a type 'g' effect on cannabis possession. Type 'a' shifts might occur, for example, when actions against poll tax default are concentrated on hotbeds of resistance, allowing a much looser enforcement in the Tory shires. Speculatively, attempts to prevent domestic murder may lead to more murders on the street by men judicially excluded from their homes.

THE TASK AHEAD

What has been presented here has been a liberalization of the concept of crime displacement, its substitution by the concepts of deflection, and more generally crime flux, and the specification of some approaches making for a potentially greater engagement of the concept of crime flux with operational police and policy decisions. It should be acknowledged that some complexities are minimized in this chapter. As an agenda towards action, we would see the following as essential elements, in this order.

1 Construction of crime impact measures involving the components of harm and consequences (but excluding crime fear). This composite measure would be the measure of crime harm incorporated in the later modelling.
2 Establishing prevalence/concentration surfaces, as described in this chapter, and testing the typology set out, using data from all three British Crime Surveys, to get a sense of time changes between the three surveys, as well as those within each survey.
3 Recalculate prevalence/concentration surfaces incorporating repeat victimizations in the measurement of the concentration dimension.
4 Test on the relative social desirability of particular crime profiles, in parallel with the next point.
5 Exploration of the predictability of crime flux by serving police-officers, using data from well-researched change initiatives.
6 Action research to determine the effects of a police resource allocation and command and control system geared to crime flux considerations.
7 Retire, exhausted.

ANTICIPATED CRITICISMS

We anticipate two types of criticism to the approach advocated. The first that it is politically naive to expect policy to be framed with explicit reference to the distribution rather than the extent of crime. One senior civil servant, on becoming acquainted with our ideas, said that she couldn't possibly go to ministers with such ideas, because of the implication that crimes were deliberately moved from one place to another. There are two elements to our rejoinder to that criticism.

First, we do not believe that displacement operates on a one-for-one basis, hence the concept of crime flux is preferred. It is not just a case of shuffling crimes around. Such a characterization is a grotesque caricature. Some reduction of serious crimes is implied in the argument.

Second, current social arrangements already determine crime placement – for example the availability of insurance determines the prevalence of insurance fraud, the income distribution the prevalence of income tax and social security fraud. If a policy consciously to redistribute crime is regarded as immoral, then the immorality must lie in the consciousness rather than the effect, since virtually all policy has an effect on the distribution of crime.

The second criticism is that what one forgoes in the substitution of the concept of displacement by that of crime flux is a clear attribution of cause. In other words, crime flux over time is not readily attributable to particular prior actions. This is true. However, we would contend that the advantage of displacement in attributing cause is spurious. The effect of changes in social arrangements on crime are not one-to-one. They are complex. They change the availability and attractiveness of criminal opportunities, the amount of time spent in different pursuits, and the relationships which might cause friction. Inferences from repeated analyses of crime flux, hopefully allied with grass-roots experience, can offer a much more realistic perspective on which policy could be based. They will never be simple, but at least an approach based on crime flux is not blinkered by choice, as has been so much of the conventional literature on displacement. Honourable exceptions are the work of Brantingham and Brantingham (1984) and Clarke and Cornish (1988). Both of these approaches are consistent with the conceptual framework of crime flux.

REFERENCES

Barr, R. and Pease, K. (1990) 'Crime displacement and placement', in M. Tonry and N. Morris (eds) *Crime and Justice*, Vol. 12, Chicago: University of Chicago Press.

Bottomley, A.K. and Pease, K. (1986) *Crime and Punishment: Interpreting the Data*, Milton Keynes: Open University Press.

Brantingham, P.L. and Brantingham, P.J. (1984) 'Burglary mobility and crime prevention planning', in R.V.G. Clarke and T.J. Hope (eds) *Coping with Burglary*, Lancaster: Kluwer-Nijhoff.

CACI (1988) *ACORN Users Guide*, London: CACI Market Analysis.

Chaiken, J.J., Lawless, M.W., and Stevenson, K. (1974) *Impact of Police Activity on the New York Subway System (Report No. R-1424-NYC)*, Santa Monica, Calif.: Rand Corporation.

Clarke, M. and Holm, E. (1987) 'Microsimulation methods in spatial analysis and planning', *Geografiska Annaler* B, 69: 145–64.

Clarke, M. and Wilson, A.G. (1987) 'Towards an applicable human geography – some observations', *Environment and Planning* A, 19: 1525–41.

Clarke, R.V.G. and Cornish, D.B. (1988) 'Crime specialisation, crime displacement and rational choice theory', in H. Wegener *et al.*, *Criminal Behavior and the Justice System: Psychological Perspectives*, New York: Springer-Verlag.

Clarke, R.V.G. and Mayhew, P.M. (1987) 'The British gas suicide story and its criminological implications', in M. Tonry and N. Morris (eds) *Crime and Justice: An Annual Review of Research*, Vol. 10, Chicago: University of Chicago Press.

Gottfredson, M. (1986) *Victims of Crime: The Dimensions of Risk*, Home Office Research Study 86, London: HMSO.

Hakim, S. and Rengert, G.F. (1981) *Crime Spillover*, Beverly Hills, Calif.: Sage.

Jones, T., Young, J. and McLean, B. (1986) *The Islington Crime Survey*, London: Tavistock.

McIver, J.P. (1980) 'External effects and the organisation of policing in metropolitan areas', paper given at the American Society of Criminology, November.

Mayhew, P.M., Clarke, R.V.G., Sturman, A. and Hough, J.M. (1976) *Crime as Opportunity*, Home Office Research Study 34, London: HMSO.

Mayhew, P.M., Clarke, R.V.G. and Elliott, D. (forthcoming) 'Motorcycle theft, helmet legislation and displacement' *Howard Journal of Criminal Justice*.

Skogan, W. (1990) 'Redesigning the National Crime Survey', *Public Opinion Quarterly* Summer.

Sparks, R.F. (1981) 'Multiple victimisation, evidence, theory and future research', *Journal of Criminal Law and Criminology* 72: 762–78.

Stenzel, W.W. (1977) 'Saint Louis High Impact Crime Displacement Study', paper given at the National Conference on Criminal Justice, February.

11

POLICING THE UPPER WORLD
Towards the global village
Michael Levi

Studies of crime, whether based on official data or victimization surveys, indicate that most crime is local, intra-class and intra-race (Herbert and Evans 1989; Jones *et al.* 1986). Both police inputs (recorded and detected crimes) and court outputs (convicted offenders) convey the overwhelming impression that whatever victims' perceptions or media accounts of 'its' seriousness (Hough and Mayhew 1985; Pease 1988), crime in England and Wales is typically small in scale, for little is stolen and physical harm is modest (Mayhew *et al.* 1989). *Inter alia*, this research might lead us to conclude that it is grossly misleading to speak of 'the' crime problem. Rather, different crimes create different problems – of actual or feared victimization – for different sectors of the population.

But there is another world of crime – not so well represented either in police or criminological accounts – that calls into question this dominant localism and smallness of scale: white-collar crime. This has caused controversy ever since the coining of the phrase fifty years ago (Sutherland 1983), and has been used to describe elite 'misconduct' from embezzlement to health and safety at work violations to the corruption of defence purchasers to the marketing by multinational corporations of infant formula food in the Third World (Levi 1987; 1989). Although the activities about which Sutherland (and many of his immediate disciples such as Clinard and Cressey) wrote were indigenous to the United States, they were not local but rather national in character – as may be evidenced by their being usually federal rather than state crimes and/or being the province of federal administrative agencies such as the Anti-trust Division, US Postal Service, and Internal Revenue Service. Indeed, at a conceptual level, the geography of where white-collar crimes occur raises difficulties largely avoided – or, more precisely, unconsidered – by ecological criminologists. Were they conceived and ordered from the inner-city boardroom, or did the planning take place at least partly in places – such as Nevada or Atlantic City casinos and bordellos, or the bathrooms and libraries of country houses – which fall well outside the urban concentric zone maps of the Chicago School but within easy range of the corporate rooftop helipad? When, after discussions between the conspirators in Abu Dhabi, the office in Aberdeen instructed the bank in Glasgow to wire the money to a bank in Switzerland, where was the crime? Where do their effects fall? The complex Contragate deals in which moneys from the sale of arms to Iran were funnelled to the Contra rebels affected Nicaragua, Iran, and Wall Street brokerage offices, since corporate share dealings are influenced by country risk decisions on the part of stock analysts. Although many of the subset of crimes that one might term planned crimes are conceived other than where they are executed, and their proceeds are 'fence' to an intermediary and/or a final consumer somewhere else again, the kinds of questions raised for geographers by such global issues in respect of white-collar crime are subtle and difficult, going beyond the normal problem of identifying the relationship between area of offender residence and crime commission (Herbert 1982; Baldwin and Bottoms 1976).

This is not merely an academic problem for geographers of crime; it goes to the heart of criminal jurisdiction, particularly under English law, and to the *policing* of fraud. Currently, none of the participants in a swindle can be prosecuted in England and Wales unless the *last* event which makes up the underlying crime takes place there. In cases that are detected before they have been successful, the swindlers are unlikely to be prosecuted overseas, and because the underlying crime was not completed in England and Wales, they cannot be tried there. In short, they will get away with it. The Law Commission (1989) report on *Jurisdiction over Offences of Fraud and Dishonesty with a Foreign Element* – to which this author gave evidence – gives as an example that

a rogue whose sole residence is in England induces his next-door neighbour, by deception, to part with a sum of money which, as it happens, is handed over in Calais when they are travelling together on a day-trip to France. The rogue cannot be prosecuted in England for obtaining property by deception, because the money was not obtained here. In the converse situation, however, the rogue and his victim being neighbours in Paris and the money paid in Dover, the miscreant can be tried here for that offence.

(Law Commission 1989: 4)

There is no jurisdiction unless the property or pecuniary advantage was 'obtained' here, and that is not an easy question to resolve.

The normal mode of policing and prosecution reflects and – via reporting rates and detection methods – reproduces the localized patterns of crime found in crime and (usually juvenile) offender surveys. The development of police forces tends to occur in a local, regional, and national context and countries exhibit a high degree of national autonomy in their setting of policing priorities. Before turning to the theme of internationalization, I shall examine the separate development of the policing and prosecution of fraud in the USA and in England and Wales.

THE POLICING AND PROSECUTION OF FRAUD IN THE UNITED STATES

Prior to the 1960s, and more specifically Watergate, the Federal Bureau of Investigation and the US Department of Justice generally took little interest in white-collar crime. The Secret Service – which has responsibility for counterfeiting and the integrity of the currency – was concerned about its sphere of interest, but this was split off from the attack on white-collar crime. Securities frauds and their prevention were the province of the Securities and Exchange Commission (SEC), established during the 1930s, and they too were not seen as a 'police problem' but rather as a 'regulatory problem'. The Federal police anyway had somewhat limited legal jurisdiction (Marx 1988) and interpreted their role under J. Edgar Hoover as primarily that of dealing with what they saw as communist subversion, the fight against 'organized crime' (crime organized by members of the *under*world) coming a poor second, and the combating of elite crime almost nowhere. However, in the aftermath of Watergate, this began to change. In particular, as Marx (1988: 39–40) observes, three high-level committees recommended expansion of activity over white-collar crime and corruption: the Attorney-General's Committee on White-Collar Crime; a Justice Department committee into the FBI's priorities and guidelines, headed by Harold R. Tyler, which suggested a shift in emphasis

219

from automobile theft and bank robbery to organized crime, white-collar crime, and political corruption; and a report by the House of Representatives subcommittee charged with FBI oversight, which repeated these views and criticized the FBI for not making more extensive use of undercover operations and complex investigations. My sources state that this was also accompanied by a shift in patterns of recruitment, favouring the recruitment of accountancy graduates who are more readily attuned to the process of financial investigations.

A different but significant dimension of the US police interest in business came via the focus on drug trafficking during the 1980s. This gradually led to the belief that if financial institutions in the USA and elsewhere in the world could be prevented from laundering the proceeds of narcotics sales, the drug problem in the USA would vanish. A more intensive drive against banks that failed to report currency deposits and transfers was mounted, leading to the conviction of many US banks (Walter 1989; Levi, 1991a). The combined effect of these factors certainly led to a substantial expansion of investigations into the business world, particularly the use of undercover tactics such as agents pretending to be gangsters depositing large suitcases full of currency which were not reported. Thus, subject to internal FBI criteria about prior suspicion of targets and to external court-imposed criteria regarding entrapment, the Sting became a tactic used against politicians and business-people, as well as against street criminals: the most famous example being ABSCAM, in which an FBI agent pretended to be a wealthy Arab wishing to set up in the United States, and leading politicians solicited and accepted bribes to assist him (and their constituents, who were to obtain jobs) (see further Greene 1981; Marx 1988).

A different sort of lesson to be learned from these changes is that it is plainly misleading to restrict 'policing' to 'the police'. The Treasury and Justice Departments co-operated from June 1980 in Operation Greenback, which used currency reporting rules to require individuals to explain where they got their cash. This merged with the Miami Organized Crime Drug Enforcement Task Forces in 1984, by which time its efforts led to indictments against 215 individuals from 82 different organizations; to the seizure of $38.8 million in suspect currency and $14.6 million in property; and to fines and taxes totalling over $2.8 billion. Operation Tracer in 1984–5 seized $100 million in alleged drug money and involved the arrest of prominent bankers.

In the USA police and prosecutors work far more closely (and under far greater prosecutorial control) than either was or even is the case – post Crown Prosecution Service – in England and Wales. Consequently, prosecutorial attitudes make a more *immediate* difference to investigations there. As Katz (1980) pointed out, following Watergate and in part caused by the moral outrage at it, there was a Social Movement against white-collar crime. Indeed, in his brilliant novel, *Bonfire of the Vanities*, Tom Wolfe satirizes the mentality of the local New York prosecutors who longed to have the opportunity to prosecute a major pillar of the Establishment instead of confining their work to what they viewed as the ethnic minority dross. This Justice for All motif was intermingled with the desire for fame and even for financial and/or political advancement that accompanied success against corporate criminals. Similar observations could be made regarding the motivation of some *Federal* prosecutors, as reported not only by Katz (1980) but also in *The Independent* (31 August 1989) in relation to Leona Helmsley, the hotel chain owner convicted of evading $1.2 million in taxes:

> It was no coincidence that the first prosecutor in the case was a former US Attorney, Rudolph Giuliani, who is now running for Mayor of New York. Yesterday's guilty verdict is expected to boost his campaign to unseat Mayor Ed Koch. Slick television advertisements telling liberal New Yorkers how a Republican candidate brought down a billionaire tax evader are to be screened in the two remaining weeks before the election primary.

(In fact, Giuliani's candidacy against a Black Democrat failed, but the case, alongside his previous successful prosecutions of major white-collar and organized crime figures, did him some political good.)

The Revolving Door system of public/private sector interchange that characterizes the US civil service generally is present in the prosecution of white-collar crime to a high degree (Katz 1980; Mann 1985). We can see this in the handling of the Dennis Levine insider dealing case in 1985, in which Bank Leu International's attorneys – where Levine's Bahamian accounts were held – were major ex-prosecutors (Frantz 1989). More recently Gary Lynch, aged 39, retired in August 1989 as director of enforcement at the SEC, to practise securities law for the law firm of Davis, Polk, and Wardwell. Interviewed by the *Wall St Journal* (28 July 1989), he stated that the one area where there was still a major regulatory problem was 'the penny stock abuses. It's all over various regions of the country'. Asked if he had any qualms about going over to the other side, he replied:

No, not at this point anyway. I think it's fair to say that I would like to have the luxury of being able to make a cut as to some types of persons that I would defend and others that I would prefer not to. There are people or institutions that are legitimate, that have a real business, and who step over the line, and have to be held accountable for it. Then there are people or institutions whose business is corrupt from the inception. It's the latter category that I'd just as soon not represent.

This author is sceptical that Lynch will have that luxury, and it is not always self-evident when a business 'is corrupt from its inception'.

Whatever motive-mongering one wishes to indulge in, it remains the case that, with the prosecution and conviction of leading figures, first, for securities fraud such as insider trading (most notably Ivan Boesky and of Mike Milken, the 'Junk Bond King', who formerly headed Drexel Burnham Lambert), and second, for currency reporting violations (most notably the Bank of Boston, Crocker Bank, E.F. Hutton, Shearson Lehmann, and the Bank of Credit and Commerce International) the last of these closed down in 1991, worldwide, US Federal prosecutors have adopted a dynamic approach in targeting upmarket white-collar criminals. Questions remain about their willingness to take on the 'real' Establishment insiders – for example over Contragate – but questions remain about whether such insiders have really been involved in crime! It appears that US police and prosecutors generally adopt an individualistic line in furthering their professional reputations, on the basis that their skills in batting for whichever team they play will make them employable and/or electable afterwards.

THE POLICING AND PROSECUTION OF FRAUD IN ENGLAND AND WALES

The early history of fraud policing in England and Wales bore analogy with that in the United States, except inasmuch as criminal legislation against insider dealing took almost fifty years longer to arrive in the UK, where it first appeared in the Companies Act 1980, subsequently amended by the Company Securities (Insider Dealing) Act 1985, the Financial Services Act 1986, and the Companies Act 1989.

Prior to the Second World War, interest and expertise in the policing of fraud was largely a matter of chance; major international bankruptcy frauds were investigated – and their perpetrators pursued in Europe and elsewhere – during the nineteenth century (Levi 1981). But the force *organization* was not such as to make that an effective method of dealing with fraud. To some extent, such an unsystematic approach to the policing of fraud may have been the product of rivalries between City of London Police and Metropolitan Police Commissioners. But it should not be forgotten that – Special Branch excepted – nation-wide policing strategies, however effectively or ineffectively implemented, are largely the creation of the 1980s. However, though Scotland Yard set up a small Sharepusher and Confidence Trickster Squad in the 1930s for frauds occurring within its jurisdiction, it was only in 1946, when the Home Office became increasingly concerned about the prospects of fraud against demobilized Britons and Americans, that a combined London Squad of twelve officers was established, entitled the Metropolitan and City of London Police Company Fraud Department. Its head has always been a Commander of the Metropolitan Police – the equivalent of an Assistant Commissioner in the City of London Police – and City of London officers, though part of the combined squad, are for disciplinary purposes answerable to the City of London Commissioner and thus effectively independent of the Metropolitan Police Commander who heads their squad. In practice, the identity of the officers from the two forces remains distinct: for promotion as well as loyalty reasons, their reference groups remain their own forces.

Although with some 588 Fraud Squad officers in England and Wales in the late 1980s, the police had more staff than the Department of Trade and Industry corporate regulation sections, they occupy what in many respects is a subsidiary role in policing the upper world. The two London police forces have approximately one-third of all the police fraud investigators in England and Wales, reflecting the domination of the metropolis in non-Scots financial services. Tracing the development of police concern about fraud, Metropolitan Police Commissioners have traditionally been more troubled about organized crime and professional criminals than about fraudsters who have no such connections: see Levi (1991b). (Sometimes, corrupt solicitors or accountants bridge the worlds of professional criminals and criminal professionals.) The City of London force gives fraud greater priority in its detective functions than does the Metropolitan force, which may reflect the relative absence of public disorder, organized crime and street crime problems in the City of London, as well as the central importance of finance capital to the prosperity of the force area. Even prior to the political and media scandal-mongering over fraud that has occurred since 1985, the City of London Commissioner repeatedly referred to the 'wide recognition' of the importance of fraud regulation and to 'public and governmental interest' in this 'crucial' area. Although total losses from fraud are far higher in the Metropolitan Police area – much greater than from other crimes in London (see Levi 1987: ch. 2) – both crime and political constituencies are much more varied in the 'Met' than in the City. Expertise in handling fraud is consequently more likely to bring one to the attention of the Commissioner of the City of London than of the Metropolitan Police. Senior officer interest in fraud *may* have increased since the mid-1980s, due to greater attention to it by television, radio, and in the main crime and news sections of newspapers as well as in the business sections that non-specialist officers are less likely to read. Moreover, as in the United States, opportunities for former police-officers in the financial world are both lucrative and considerable, making expertise in financial crime investigation a more desirable attribute than has been the case in the past.

There are two reasons why the policing of major frauds might become tougher in the future. The first relates to the draconian powers of investigation granted to the Department of Trade and Industry inspectors under sections 177 and 178 of the Financial Services Act 1986, which, in cases of suspected insider dealing, enable inspectors to require information from any persons on pain of imprisonment for contempt of court and/or de-authorization from conducting investment business. (These powers are dealt with in Levi 1987; 1991a). The second source of change in fraud policing relates to the establishment of the Serious Fraud Office (SFO) under the Criminal Justice Act 1987 (Wood 1989).

The SFO is a statutory body responsible directly to the Attorney-General, headed by a Director who is independent of the Director of Public Prosecutions, and with some eighty accountants, lawyers, investigators and clerical staff, including several partners seconded from leading firms of City accountants. Its brief is to deal with 'serious' and/or 'complex' fraud, loosely defined as frauds involving at least £2 million 'at risk' where the details are 'complex'. It is uncertain how complexity and seriousness will be (or indeed can be) defined, and section 1(3) of the Criminal Justice Act 1987 makes this remit quite open when it states that 'The Director may carry out, in conjunction with the police, investigations into any suspected offence which appears to him to involve serious or complex fraud'.

The Director of the Serious Fraud Office is given major powers to require documents and answers to questions as a result of the Act, tougher in some respects than those available under the Police and Criminal Evidence Act 1984 and almost as great as those in the Financial Services Act 1986 (see Levi 1987: chs 5 and 8; 1991a; Wood 1989). Except for revenue intelligence, which will be disclosed to others only for the purpose of a criminal prosecution by either the Serious Fraud Office or, in relation to an Inland Revenue offence, to the Crown Prosecution Service – see section 3 of the Criminal Justice Act 1987 – information obtained may be passed on not only to the police but also to Department of Trade Inspectors, the Official Receiver, and, under section 3 (6),

(c) any body having supervisory, regulatory, or disciplinary functions in relation to any profession or area of commercial activity; and

(d) any person or body having, under the law of any country or territory outside Great Britain, functions corresponding to any of the functions of any person or body mentioned ... above.

This has great possibilities for the development of international intelligence and supervisory interchange. Section 3 (4) also permits the Director to enter into agreements to supply for an (unspecified) specified purpose information that is in his possession.

After initial rapture, the Serious Fraud Office has had a difficult birth, and the media have by no means treated it as a dream-child. Although judges and juries are outside the control of even the most competent investigators and prosecutors, the acquittal of several defendants – including the alleged Lloyd's of London fraudsters, Ian Posgate and Kenneth Grob – and the lengthy delays in the inquiries into the takeover of the House of Fraser by the Al Fayeds and in the coming to trial of Guinness and Barlow Clowes defendants in 1989–90 have not given well-wishers much encouragement.

In part, this difficulty may be due to the novelty for English criminal lawyers of being part of a team, but it may also reflect the extant problems in obtaining admissible or even informative evidence from some overseas jurisdictions favoured by fraudsters. Nevertheless, even if its arrival has not been greeted with universal acclaim, it represents more than a symbolic gesture in the direction of greater priority of fraud prosecution. For the first time, there is a dedicated agency with a bureaucratic interest in prosecuting fraud successfully – its principal *raison d'être* – which does not have to wait in the Director of Public Prosecutions' queue for resources. And it has taken on – albeit largely by referral from the government – major allegations which are being pursued in a more criminal trial-oriented way than was present in some past police inquiries.

THE INTERNATIONALIZATION OF FRAUD POLICING

In 1988 there were some $1,200 *billion* in cross-border securities transactions, but even this large figure is dwarfed by foreign exchange and commodities futures transactions. Drucker (1989) notes that

> The transnational economy is shaped mainly by money flows rather than by trade in goods and services. ... The monetary and fiscal policies of sovereign national governments increasingly react to events in the transnational money and capital markets rather than actively shape them. ... increasingly decision-making power is shifting to ... the region ... there is a genuine – and almost autonomous – world economy of money, credit and investment flows. It is organized by information that no longer knows national boundaries. Finally there is the transnational enterprise ... which views the entire developed non-communist world as one market, indeed as one 'location', both to produce and to sell goods and services. ... while the transnational world is reality, it still lacks the institutions it needs. Above all, it needs transnational law.
>
> (Drucker 1989: 109–11)

Fraud reduction is not the sole motive for regulating investment markets, but how are fraud controllers reacting to this globalization? Many parallel developments in legislation against and policing of business crime are occurring, though some US practices have been followed in the UK and, with varying degrees of enthusiasm, in Japan, Hong Kong, France and the EC regionally. This UK/US parallelism applies even to the public/private sector transfers by fraud investigators (to regulatory agencies and private fraud security firms) and – less frequently – by fraud investigators and securities regulators to private sector areas such as law firms that defend white-collar accused. The latter trend may increase, just as it did in the United States in the post-Watergate period, that is since white-collar workers became much more likely to be prosecuted for their offences. As this trend increases, it may be that fraud investigators will become increasingly defensive about revealing their methods to people who may soon be working for the 'other side' (as is the case with the police *vis-à-vis* English prosecutors, who may later defend).

There remain major areas of national divergence. Few British police or prosecutors go into politics, and fraud-busting is more likely to wreck than to launch political careers in Britain. Despite the changes described here, trustworthiness in 'not rocking the boat', unless one's political masters want it rocked, remained important even in Thatcher's radical Britain, as it is also in Japan, where the scandal over the supply of political funds to the Liberal Party by Recruit Cosmos via preferential terms for new share issues led in 1989 to the resignation and prosecution of major party and business leaders.

But why is there this multinational convergence in the direction of high-profile fraud investigation and prosecution bodies? Interpretation is not wholly objective, but the convergence arises largely because of the risks that fraud is posing to popular capitalism, to investments on securities markets and to deposits in financial institutions. It is also happening because almost all serious securities frauds involve extraterritorial informational needs by national agencies, and these have to be provided for by national legislation and Memoranda of Understanding. In the USA deposit insurance has given the Federal government a financial reason (and a legal authority) to intervene in banking and in savings and loans supervision. In Britain, despite the fact that privatization share issues have led to very uneven share ownership, the greater interest and involvement in the stock market by all but the poorest sections of society has increased the *political* risks arising from fraud. So securities fraud, combined with concern about the possibility of 'organized crime' and terrorist groups (i.e. those who are not allies) laundering money through the financial markets, has shifted

227

government and police in the direction of setting up better organizational capability of analysing crime through money flows and of dealing with sophisticated financial crime.

So far, the policing of fraud has been discussed as if it were a national or multinational rather than a genuinely *inter*national issue. In England and Wales, controversy is aroused still by episodic floating of the idea of a national police or detective force, not only because of fears of a police state but also because this is regarded as an inappropriate way of dealing with small-scale local crime and public order phenomena. In his speech to the Police Superintendents' Association (26 September 1989), the Home Secretary was careful merely to *suggest* the desirability of a National Crime Intelligence organization: given the *amour propre* of Chief Constables and Police Authorities, this modesty was politically wise. Chief officers could not agree on proposals to establish a national intelligence and investigation body and the notion was referred back to the Crime Sub-committee of ACPO to examine its operational value and feasibility (*The Independent*, 19 January 1990) and subsequently in 1991, a National Criminal Intelligence – but not investigative – body, has been established. Take this one step further and discuss international policing, in the sense of officers with supranational powers, and the critics of the idea of a national force are likely to become apoplectic. The theme of international co-operation in combating organized fraud runs through the Annual Reports of the Commissioner of the Metropolis during the 1970s and 1980s, for instance in 1974, 1975 and 1976, but except at an informal level, remains haphazard in practice. International police harmonization raises important constitutional and accountability questions: as Anderson (1989) observes, criminal law is based upon concepts of national sovereignty, and this imposes critical restraints.

Policing powers are founded upon law and the discretion allowed by the visibility of police actions and the tolerance of the courts. Coping with overseas jurisdictions, the difficulty of informal as well as formal action increases considerably, since the patterns of pressure, friendship, Masonic Lodge membership, and so on make for fewer and less powerful trade-offs and for greater uncertainty about responses in different cultures. Secrecy statutes also present legal obstacles to the conveying of information which many UK police officers – used to the informal methods of the Ways and Means Act – find particularly irksome (Levi, forthcoming).

There is evidence of a greater willingness on the part of the

English law-makers to facilitate (1) extradition (Extradition Act 1989); (2) jurisdiction over international crime; and (3) mutual judicial and extra-judicial assistance. Some of these changes affect the police, but others affect both the Serious Fraud Office and the plethora of 'Self-Regulatory' Organizations established under the Financial Services Act 1986 who are authorized by the Securities and Investments Board, which in turn is answerable to the Department of Trade and Industry. The regulation of financial crime has largely been ignored by criminologists who are preoccupied with studying the policing of the working class and lumpenproletariat, which remains (and will remain) the dominant mode of police (and policing) activity.

These developments should be seen within the growing trend of international regulators – of which the principal UK agency is the Department of Trade and Industry – to conclude information-sharing agreements, normally bilaterally (as in most US treaties), but sometimes under the aegis of the Bank for International Settlements, the Basle Committee of banking regulators, the Commonwealth, the European Economic Community or the OECD (1989) who have agreed to fairly routine exchange of previously sacrosanct revenue information. Authorized UK regulators such as the Securities and Futures Association can enter into information-sharing arrangements direct with their counterparts overseas, which speeds up the communications process by side-stepping the wait for central agencies – which each have their own priorities and have a less direct stake in outcomes than do the agencies that want to *use* the information; such agency-to-agency agreements enhance the informal network still further. In 1988 the US Securities and Exchange Commission was granted the power to collect information on behalf of overseas regulators and to pass it on to them. The Companies Act 1989 gives the UK Department of Trade and Industry the power to obtain under compulsion information on behalf of overseas regulators. The Global Village aspect of this is derived from these international trends.

Nation-states may exercise *some* control over business and over some types of business crime, but co-ordination is necessary whether in respect of transfer pricing between the vertical and horizontally integrated sections of multinational corporations to avoid tax, or of the choice by securities traders among London, New York or Tokyo, the spatial and time zonal relationships of which are central to international finance capital. Anyone believing that securities markets are nationally independent need only look at the mutual involvement of these exchanges in the stock market crashes of October 1987 and 1989. The need for compatibility, if not complete harmonization, of rules and their application is particularly great once – as in financial services – one accepts that authorization by other countries and/or regulatory authorities gives the right of access to one's own markets. The move towards harmonization of banking and securities trading in Europe before 1993 will necessitate total mutual acceptance within the EC of banks established in other EC countries – so that there no longer will be any British Banks as such – and also Britain's acceptance as UK dealers of firms authorized to sell securities on, for example, the Milan stock exchange: an exchange with a more relaxed approach to regulation. Likewise, at least until recently, UK securities regulation was much looser than that in the US, and remains so in respect of disclosure requirements before a stock is listed on the market: an indicator of this is the choice by the ill-fated ISC defence corporation to be listed in London rather than in the USA, allegedly specifically to avoid the disclosure requirements imposed by the Securities and Exchange Commission. Ferranti's subsequent purchase of this business led to its losing some £215 million, allegedly by fraud discovered in 1989. The Bank of Credit and Commerce International allegedly headquartered in Luxembourg and the Cayman Islands *because* their regulations could not cope. The political choice between regulatory standards therefore is consequential in fraud terms as well as in enterprise terms. The growth of cross-border securities flotations has led the International Organization of Securities Commissions – founded only in 1987 – to begin in 1989 the difficult process of co-ordinating rules and regulatory practice, the objective being, *inter alia*, to permit issuers of securities to prepare a single disclosure document for use in each jurisdiction in which it proposes to issue securities. This would save enormous time and cost for companies who wish to sell equities in multiple jurisdictions.

The Global Village concept implies harmony of objectives. That is always a problem with the political usage of the word 'community' to conceal underlying conflicts of interest by 'talking the market up' into the appearance of harmony, thereby perhaps producing a halo effect whereby there actually is less dissent. In the world of serious money, financial services suppliers and governments have a powerful collective interest in maintaining public confidence in the legitimacy of their markets. But beyond that it is Hobbes' *bellum omnium contra omnes* and as in the banking secrecy and tax haven businesses (Levi, 1991a), many smaller jurisdictions feel that they can survive only by offering less regulation, particularly if the political and financial costs of such non-regulation fall upon others. Apart from any belief in the intrinsic superiority of US law and institutions, the attempt to produce a global-level playing field of securities rules is one method of reducing the comparative disadvantage that US traders feel results from their own traditionally highly regulated domestic markets: whose rules themselves were the product of earlier market excesses in the 1920s. The internationalization of fraud regulation is not explicable totally in terms of economic self-interest of the large players in squeezing out the marginal ones, but as in the case of health and safety legislation during the nineteenth century (Carson 1974), enlightenment prospers best when allied with self-interest.

REFERENCES

Anderson, M. (1989) *Policing the world: Interpol and the Politics of International Police Co-operation*, Oxford: Clarendon Press.

Baldwin, J. and Bottoms, A.E. (1976) *The Urban Criminal*, London: Tavistock.

Carson, W.G. (1974) 'Symbolic and instrumental dimensions of early factory legislation', in R. Hood (ed.) *Crime, Criminology, and Public Policy*, London: Heinemann.

Drucker, P. (1989) *The New Realities*, London: Heinemann.

Frantz, D. (1989) *Mr. Diamond*, London: Pan.

Greene, R. (1981) *The Sting Man*, New York: Elsevier-Dutton.

Herbert, D.T. (1982) *The Geography of Urban Crime*, London: Longman.

Herbert, D.T. and Evans, D.J. (eds) (1989) *The Geography of Crime*, London: Routledge.

Hough, M. and Mayhew, P. (1985) *Taking Account of Crime: Key Findings from the British Crime Survey*, London: HMSO.

Jones, T., Maclean, B., and Young, J. (1986) *The Islington Crime Survey*, Aldershot: Gower.

Katz, J. (1980) 'The social movement against white-collar crime', in E. Bittner and S. Messenger (eds) *Criminology Review Yearbook*, Vol. 2, Beverly Hills, Calif.: Sage.

Law Commission (1989) *Jurisdiction over Offences of Fraud and Dishonesty with a Foreign Element*, London: HMSO.

Levi, M. (1981) *The Phantom Capitalists: The Organisation and Control of Long-Term Fraud*, Aldershot: Gower.

—— (1987) *Regulating Fraud: White-Collar Crime and the Criminal Process*, London: Routledge.

—— (1989) 'Recent texts on white-collar crime: an overview', *British Journal of Criminology* 29 (4): 412–15.

—— (1991b) 'Fraudbusting in London: developments in the policing of white-collar crime', in S. Whimster and L. Budd (eds) *The City of London*, London: Routledge.

—— (1991a) *Customer Confidentiality, Money-Laundering, and Police–Bank Relationships*, English Law and Practice in a Global Environment, London: Police Foundation.

Mann, K. (1985) *Defending White-Collar Crime: A Portrait of Attorneys at Work*, London: Yale University Press.

Marx, G. (1988) *Undercover: Police Surveillance in America*, London: University of California Press.

Mayhew, P., Elliott, D. and Dowds, L. (1989) *The 1988 British Crime Survey*, London: HMSO.

OECD (1989) *Explanatory Report on the Convention on Mutual Administrative Assistance in Tax Matters*, Strasbourg: Council of Europe.

Pease, K. (1988) *Judgments of Crime Seriousness: Evidence from the 1984 British Crime Survey*, Home Office Research and Planning Unit Paper 44, London: Home Office.

Sutherland, E. (1983) *White-Collar Crime: The Uncut Version*, London: Yale University Press.

Walter, I. (1989) *Secret Money*, 2nd edn, London: Unwin Hyman.

Wood, J. (1989) 'The Serious Fraud Office', *Criminal Law Review*, March: 175–84.

12

POLICE PRACTICES AND CRIME RATES IN THE LOWER WORLD

Prostitution in Vancouver

John Lowman

Despite a few passing (and usually unheeded) cautions about the problems involved with interpreting crime statistics, geographers have generally proceeded from the assumption that police-generated crime statistics reflect the behaviour of offenders as opposed to the actions of the people constructing the statistics. Until quite recently they largely ignored the constructionist or institutionalist perspective (cf. Biderman and Reiss 1967; Kitsuse and Cicourel 1963) in sociology and criminology which, in its most extreme rendition (e.g. Ditton 1979), suggests that crime statistics can be understood only in terms of the activities of the people they are produced by rather than those they are supposed to be about. From a geographic viewpoint one of the main implications of this perspective is that rather than providing a literal portrait of the geographic and social distribution of crime, official statistics map the selective application of the law by police and other agencies; they are a matter of ascription, not description, or, at least, if they do describe something it is not what common-sense suggests. Cicourel (1968) stated the case this way:

My observations suggest police and probation perspectives follow community typifications in organizing the city into areas where they expect to receive the most difficulty from deviant or difficult elements, to areas where little trouble is expected, and where more care should be taken dealing with the populace because of socio-economic and political influence. The partition of the city into areas of more or less

233

anticipated crime provides both police and probation officers with additional typifications about what to expect when patrolling or making calls in the areas. Thus the officer's preconstituted typifications and stock of knowledge at hand leads him to prejudge much of what he encounters. ... Thus particular ecological settings, populated by persons with 'known' styles of dress and physical appearance, provide the officer with quick inferences about 'what is going on' although not based on factual-type material he must describe sooner or later in oral or written form.

(Cicourel 1968:67)

While this description may capture well the mind set of the line officer as he patrols the city, subsequent research in Britain and North America has forcefully suggested that *intra*-urban crime rates cannot be interpreted simply as artefacts of police activities because the majority of criminal events which come to the notice of the police are reported to them by members of the public (Bottomley and Coleman 1981; Mawby 1979; McCabe and Sutcliffe 1978). Numerous observational studies of the police suggest that they do select certain types of person for scrutiny, but that most police interactions with citizens do not end up with the 'discovery' of a crime (see e.g. Ericson 1982). Thus while writers like Matza (1969) suggested that police utilize 'methodical' rather than 'incidental' suspicion when enforcing criminal laws,[1] he, in particular, did not recognize the central role of members of the public in identifying both crimes and suspects. As Bottomley and Coleman (1981) point out, while methodic suspicion and what Sacks (1972) referred to as 'incongruity procedures' may play an important role in certain types of 'proactive' police work,[2] and guide the process by which patrol police generally 'order the streets' – often without needing to lay criminal charges (Ericson 1982) – such arguments confuse methods of suspicion with methods of detection (see also Mawby 1979). To the extent that much police work which results in the laying of criminal charges is a reaction to citizen complaints, the method of police suspicion would appear to be less important in the process of identifying criminals than is the information provided by complainants. Of course, this still leaves the possibility that crime statistics tell us as much about public images of crime as they do about 'criminal' behaviour, that police activity can importantly influence the nature of crime reports flowing from the public, and that police practices can help to consolidate 'outlaw' identities. But more of these issues later.

There can be no doubt that one of the most important studies in the unfolding debate about the interpretation of police-generated crime statistics was Rob Mawby's *Policing the City* (1979), the second monograph emanating from an extensive study of crime and policing in Sheffield. But important though it may have been as a corrective to the idea that crime statistics are purely an artefact of police practices, the question remains as to whether this study was able to provide definitive conclusions about all the propositions that it aspired to test, or whether it tested all the propositions that are relevant to an understanding of the effects of police practices on intra-urban crime rates. Elsewhere (Lowman 1982) I have suggested that it was limited in both respects. The purpose of these observations was not so much to 'defend' Cicourel and Matza (Mawby 1989: 267), but to urge geographers to engage in debates raging in criminology over issues which threatened to make complete nonsense of the geography of crime. In the process of introducing these debates to geographers I suggested that, while there was compelling evidence to suggest that crime statistics could not be written of as an artefact of police practices, it would be misleading to interpret them as not being influenced by police practices at all.

In suggesting that this view once again requires qualification I am going to argue that the effects of police practices on local problems and the relative impact of police activities on crime rates are related and that the foreclosure of certain theoretical and empirical issues is unwarranted. A discussion of prostitution policing in Vancouver, Canada, will be used to illustrate the contention that we should remain mindful of the influence of police practices on both the location of offences and on who is processed as criminal, especially when it comes to proactively policed *targetless* crimes.[3] It should be noted at the outset that this example is not meant to suggest that intra-jurisdictional crime rates can be interpreted only (or, in the cases of some other kinds of crime data, mainly) as artefacts of police practices; this example should not be carelessly generalized. But a review of some of the debates still raging in the criminological literature will nevertheless suggest that, in a very general sense, crime rates cannot be interpreted in isolation from the context of policing and that we should view with suspicion any transhistorical and transcultural conclusion in this respect.

INTER-JURISDICTIONAL CRIME PATTERNS AND TEMPORAL TRENDS

In terms of the geography of crime it is worth noting in passing that Mawby's comments about police practices and crime rates are restricted to analyses involving a single police jurisdiction. When it comes to inter-jurisdictional studies he acknowledges that differences in policing and crime recording can have a profound influence on crime rates. One classic example is provided by Farrington and Dowds (1985) in their study of the factors explaining differences in crime rates in three English counties, and many other examples have been described in the literature (cf. O'Brien 1985: 35, 102–5; Bottomley and Pease 1986: 34–41). Although such examples have led some authors to question the utility of Uniform Crime Reports (UCRs) for inter-jurisdictional analyses, others have suggested that at least for certain index crimes, UCRs and victimization surveys reveal similar crime patterns and offender profiles. Perhaps the most important paper in this genre is by Gove et al. (1985) who, from a comparison of UCR and National Crime Survey (NCS) data in the USA, conclude that UCR data on motor vehicle theft, robbery and burglary provide a reasonably reliable index of violations (as is the case with homicide too, although obviously no comparison could be made with victim survey data in this respect). In contrast, they say no such conclusion could be drawn with larceny, hence their suggestion that analysis of aggregate crime rates should be avoided since larceny constitutes such a large proportion of those rates. They also argue that UCRs provide more reliable information about rape and aggravated assault than do victim surveys. When it comes to motor vehicle theft and robbery certainly the correlations of UCR and NCS data (Gove et al. 1985: 478) are quite high although even some of these coefficients leave a substantial amount of variance unexplained. One of the main points of Gove, et al. analysis is to establish the reliability of UCRs in the case of more 'serious' crimes. But even then it should be remembered that these categories constitute only a small proportion of 'crime'.[4] And their conclusions beg all sorts of questions about what crime 'seriousness' might mean, especially when one considers that members of the public can report only crime victimizations of which they are aware – consumer fraud, price fixing, infractions of worker safety regulations, pollution law violations, and so on, are, by implication, all rendered 'non-serious' in an analysis which 'suggests that the UCRs appear to reflect fairly accurately what the citizens and police perceive as to violations of the law which pose a significant threat to social order' (Gove *et al.* 1985: 490).

When it comes to time series analysis, there are many examples of the temporal effects of changes in police recording practices on crime rates (for examples see Bottomley and Coleman 1981: 81–5). In this respect Uniform Crime Reports and victimization surveys give quite different impressions of certain crime trends. In Britain, for example, the increases in rates of burglary indicated by police data between 1972 and 1980 are not reproduced by victim surveys (Bottomley and Pease 1986: 22–3). The same divergence between UCRs and National Crime Survey data is apparent in the USA for burglary and several other types of crime (US Bureau of Justice Statistics 1988). And if Mawby's cross-sectional analysis produced some faith in the degree to which crime statistics reflect geographic variations in offence patterns rather than police practices, Pepinsky's (1987) study of temporal trends in crime rates in the same city only a few years later yielded exactly the opposite conclusion.

While the reasons for the divergences between crime rate trends as measured by UCRs and victim surveys are still the subject of considerable debate, the point of mentioning them here is to suggest that although the role of the police in the criminal labelling process has not turned out to be as crucial as some authors speculated, there is much more to understanding the effect of police practices on crime rates than this (even if the discussion is restricted to intra-jurisdictional crime patterns). Although the police response to crime is often citizen initiated, this does not mean to say that crime rates unproblematically reflect offender behaviour; public reporting behaviour may also significantly influence such rates. In this regard, the US National Crime Surveys from 1973 to 1987 indicated that the percentage of crimes reported to the police has 'increased significantly since the survey first began in 1973' (US Bureau of Justice Statistics 1988). The overall number of victimizations (including those not reported to the police) increased from 1973 to 1981, but then gradually fell so that in the period 1985, 1986 and 1987 the numbers were lower than in any of the preceding twelve years. In contrast, UCR rates steadily increased from 1973 to 1987.

INTRA-JURISDICTIONAL CRIME RATES

In examining the implications of *Policing the City* for interpreting intra-jurisdictional crime rates there can be no doubt that Mawby presents convincing empirical evidence to show that certain kinds of offender residence and offence patterns at certain points in time reflect offender behaviour as much or more than they do police practices. Nevertheless, this particular study was limited in several ways so that generalization from it is difficult:

1 since eight of the nine study areas were described as 'working class' it did not provide a way of testing hypotheses about police practices in relation to different social classes
2 similarly it did not provide a means of testing hypotheses about police discretion and the 'race' of offenders (a key issue for American writers)
3 it did provide some evidence of the impact of police practices on certain kinds of crime rate (his findings applied mainly to indictable crimes)
4 it did not offer a historical or processual perspective on the amplification of deviance hypothesis.

(Lowman 1982)

In reflecting on these observations Mawby (1989) did not comment on the first two points. As to the third, he apparently interpreted it as suggesting that *Policing the City* dealt only with indictable crime. This would indeed be a 'mysterious' reading of the book, as he puts it, given its extensive reference to non-indictable crime. But this comment was meant to suggest that his general finding – that crime rates are not influenced by police practices – was not fully supported by some of the evidence provided about non-indictable crime.

In response to the fourth point Mawby notes that the Sheffield study was not entirely ahistorical to the extent that another publication emanating from it (Bottoms and Xanthos 1981) provides evidence that local authority housing policies and 'tenant selection/choices' play a role in creating 'crime- (or problem-) prone environments' and in this respect 'the housing department seemed a more influential agency than the police' in the genesis of a 'criminal area'. Fair enough, but this discussion of housing policy does not actually address the argument that a cross-sectional study of policing cannot deal with certain kinds of hypothesis emanating from the *amplification of deviance* literature. One way that selective

238

law enforcement practices may play a role in creating and/or amplifying criminal behaviour is by helping to galvanize the 'outlaw' identity of certain groups and provoke certain kinds of criminality. One classic statement of this sort is provided by Lemert's (1951) distinction between 'primary' and 'secondary' deviance. The concept of deviance amplification itself was first articulated by Wilkins (1964) in an application of the cybernetic principle of *mutual causality* to the crime–control relationship. Such perspectives seek to understand the ways in which control practices can amplify crime, not just select out certain kinds of people to label as criminal. In the case of 'criminal areas' such perspectives might examine the way that certain neighbourhoods come to symbolize disreputability through a process which Brantingham and Brantingham (1981) refer to as 'ecological labelling', and might implicate police practices in that process. A cross-sectional analysis simply cannot test this kind of hypothesis.

One example of an analysis that does provide some insight into the police role in amplifying crime is appropriated by Mawby to bolster his own position. He contends that although the method employed was ethnographic, Owen Gill's *Luke Street* supports the contention that police practices do not affect crime rates. He notes that

> juveniles in the area were particularly vulnerable to arrest for minor street offences but that this tended to supplement other arrests. That is, rather than create a higher offender rate, police strategies resulted in the same offenders being arrested more frequently.
>
> (Mawby 1989: 270)

No matter how one looks at it, this police activity is certainly creating a higher *offence* rate. But Gill argues that the negative stereotyping of an area not only reflects the incidence of crime and delinquency but also is partly productive of it (Gill 1977: 62). One of the points of his argument is to suggest that the largely unstructured and public lives of unemployed youths makes them come into regular and mainly adversarial contact with police. A cross-sectional analysis cannot shed much light on the proposition that a youth's 'outlaw' identity – which might result in the commission of a subsequent offence that is reported to the police by a member of the public – is forged through his ongoing interactions with police (even though many of these interactions do not end up being defined as a 'crime'). How general this process may be is a different issue, but if it is not thought to be widespread now,[5] one should not rule out the idea that it might have been more important at some point in the past or might become much more important in the future. The important point of this example is that empirical work focusing on the criminal event – remembering that only a small amount of police time is actually taken up by interactions with citizens which end up in the laying of a formal criminal charge[6] – may not tell the whole story about the impact of police practices on crime rates. The operation of the 'sus' law in Britain provides one example of the way that police select out certain groups for attention, particularly Black people (Willis 1983), and local crime survey research suggests that what comes to be perceived as police harassment can cause considerable resentment (Jones *et al.* 1986: 145–9). If the direct impact of police practices on intra-jurisdictional crime rates is nothing more than a historical oddity one wonders what to make of Westwood's (1990) tentative finding that Home Office guidelines on cautioning (rather than prosecuting) offenders appear to 'inadvertently lead to a disproportionate prosecution rate for offenders from poor inner city areas and certain local authority areas'.[7]

TRENDS IN THEORY AND RESEARCH

Although it would seem that the emphasis of much research on the police has indeed shifted away from some of the issues that preoccupied criminology in the 1960s and 1970s, it might be unwise to conclude that they should be laid to rest. For one thing, why should we assume that the impact of police discretionary practices on crime rates will remain constant? And while researchers may have turned to examination of the 'effects of particular police practices on local problems', (Mawby 1989: 277) are these different issues mutually exclusive?

Mawby enlists left realism as one indication that criminology has moved beyond the notion that crime rates are simply an artefact of policing. Certainly left realists have taken the position that there are geographic variations in crime rates (Lea and Young 1984: 41). But is it correct to say that 'as the crime problem has been introduced as an acceptable agenda by the left, so the assumption that the spatial location of offences is an artefact of differential policing has been dismissed' (Mawby 1989: 272)? There are at least two problems with this statement. The first is the casual slippage from a discussion of left realist arguments to a reference to 'the left' as if it is a homogeneous entity (a slippage which would, no doubt, leave some British criminological leftists cringing in their boots). Left realism, it should be remembered, emerged polemically – a critique of 'left idealism' as well as the realism of the right (Lea and Young 1984). Only the most unconcerned of readings of the leftist constituency of British criminology could leave the impression that left realism has created a consensus. The second is the inference that while left realists argue that area differences in crime rates cannot be sceptically written off as an artefact of police practices, this is not the same as saying that police practices have no impact on crime rates. In this respect the words of Jock Young, doyen of left realism, are illuminating:

> The most crucial dimension of deviance is that it is a product of action and reaction, of actors and reactors, of behavior and of rules. ... Each part of this dyad is vital to realism. ... The crime rate ... must, of necessity, involve behavior and rule enforcers. ... The notion of a 'real' crime rate, independent of social reaction – a simple measure of changes in criminal behavior – is just as absurd as that of an epiphenomenal crime rate merely created by reactors. ... the differential rates of crime between different social groups cannot be merely the result of different deployment of police resources nor merely differences in behavior: they must – whatever the precise weighting between them – involve the two factors. The reductio ad absurdum of an either-or analysis is that all people behave similarly and the police act differentially between them, or that criminality varies and the police act totally similarly to every group. Either ideal scenario would be so extraordinary as to be untenable.
>
> (Young 1987: 339)

It is probably fair to say that criminology has moved on from the kinds of problems that Mawby was examining in *Policing the City*. And left realism has, through its development of the local crime survey (e.g. Jones *et al.* 1986), firmly acknowledged that there is a general symmetry of victims and offenders in terms of their social class. But on closer inspection left realism does not, as Mawby implies, adopt a one-sided prima-facie interpretation of official crime statistics. Indeed, some of the major works of left realism (Lea

241

and Young 1984; Kinsey *et al*. 1986) describe a deviancy amplification spiral in inner city neighbourhoods as being initiated by a style of policing which alienates the community from the police thereby restricting information flows from the public to the police, which in turn undermines the power of the police to do anything about crime, which in turn results in increases in crime, which in turn undermines the flow of information from the public to the police – and so on. Again this *process* is beyond the scope of a cross-sectional analysis.

Another issue that local crime surveys raise is the influence of the police on public perceptions and definitions of crime, and thus on what kinds of incidents are reported to the police as crimes (Jones *et al*. 1986; MacLean, forthcoming).[8] It is perhaps for this reason that one of the architects of the Islington Crime Survey, despite acknowledging that the types of crimes which can be measured by this technology are spatially concentrated (Jones *et al*. 1986: 83), nevertheless adopts an institutionalist perspective on crime statistics, albeit one with an unavowedly structural and political edge (MacLean, forthcoming).

As to the claim that 'the most telling rebuttal of the argument that police practices significantly influence area crime rates (or at least offence rates) comes, implicitly, from victimization surveys' (Mawby 1989: 271) again there are reasons to believe that a more cautious tone is in order. While there can be little doubt that victimization surveys indicate that offence rates do vary from city to city, between rural and urban areas and from suburb to inner city, it is one thing to say that these findings suggest that crime rates are not purely an artefact of police practices, but quite another to argue that police practices have no 'significant' impact on crime rates. The problem here is that very few victimization surveys use sample techniques and/or sizes which allow us to say very much at an intra-urban level about the correspondence between victimization survey and official crime rates. One local victimization survey which did manage to achieve such a comparison – in this case in three London boroughs – concluded that 'patterns of crime in the three areas look very different if one considers survey data rather than police statistics' (Sparks *et al*. 1977: 154). The survey showed that Brixton and Kensington had roughly the same overall crime rates as measured by victimization surveys, but the officially recorded crime rate of Brixton was only three-fifths of that found in Kensington. How much police practices, particularly record-keeping, are responsible for these differences is open to conjecture. Certainly one might like to see many more of these comparisons before reaching any firm conclusion.

There is another sense in which Mawby's reference to 'crime' and victimization surveys is problematic. Mawby's work deals only with certain kinds of crime – that for which police statistics are available. But Mawby consistently uses the term 'crime' generically without being sensitive to some of the ideological and empirical issues that such usage entails. He says nothing about the burgeoning research on white-collar, governmental and corporate infractions, species of crime which neither victimization surveys or official crime statistics tell us very much about. To the extent that many of these types of infractions have never been policed in anything more than a rudimentary way, it goes without saying that (lack of) policing has a profound impact on patterns of crime.

PROACTIVE POLICING, CRIME AREAS AND PATTERNS OF OFFENDER RESIDENCE: AN EXAMPLE

It has already been argued elsewhere (Lowman 1982: 335) that while Sheffield police might treat 'with some amusement' the idea that their activities somehow 'created' area differences in respect to prostitution (Mawby 1979: 146) such a finding might not hold in other contexts. One illustration of this point was the way in which law enforcement decisions in Vancouver in 1975 displaced prostitutes from indoor locations on to city streets and created a more intractable law enforcement problem in the process. Then, in 1978, efforts to control street prostitution were confounded by the first of a series of court decisions which are said to have emasculated the street prostitution law (the 'soliciting' law) which was in effect at that time.[9] Police attempts to deal with the street sex trade since then have thrown into even sharper relief the incautiousness of Mawby's overgeneralization from a single example.

Attempts to control street prostitution in Vancouver since this displacement have proven instructive in terms of the present discussion in two ways: first, much of the changing geography of street prostitution in Vancouver since 1980 has been the direct result of law enforcement efforts, and second, deployment of law enforcement resources importantly influences the types of offenders being processed by the criminal justice system.

'Like trying to trap mercury with the underside of a spoon'

From 1975 to 1984 there were five main prostitution strolls in Vancouver (Figure 12.1). The two most populous strolls in terms of the number of people working in them were the West End and Georgia Street. The West End was also the only one of the five areas that was primarily residential.

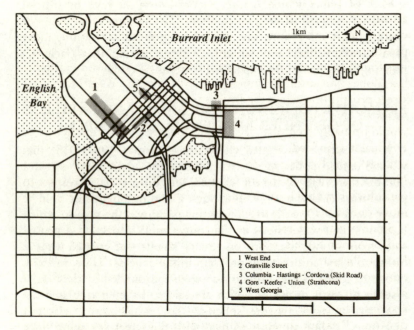

Figure 12.1   Vancouver prostitution strolls 1975–84

When it appeared that the law could no longer control street prostitution municipal and provincial government politicians, police and neighbourhood lobby groups began to vigorously campaign for the Federal government (which has exclusive responsibility for enacting criminal law) to revise the *Criminal Code*. The first campaign was initiated by commercial and business interests along Davie Street (a commercial strip and the main thoroughfare of the West End) and Georgia Street (a hotel and office district). The main police response was to say that their hands were tied by inadequate laws although they did institute two prostitution Task Forces in 1979 and 1981 in an attempt to shift prostitutes from the areas generating complaints. One effect of these police efforts may have been to force some prostitutes into the residential back streets of the West End although this 'spread' was probably accentuated by an actual increase in the number of prostitutes working in the area. In 1981 the 'Concerned Residents of the West End' (CROWE) was formed with the single purpose of ridding their neighbourhood of prostitutes and their clients.

In 1982 the Vancouver City Council enacted a by-law in an attempt to suppress street prostitution. After the laying of some 800 charges under this law (but with no impact on levels of street prostitution) it had to be scrapped when a similar law in Calgary was found to infringe on what should be the jurisdiction of criminal law alone. Not until the summer of 1984 was the West End rid of street prostitution when the British Columbia government issued a series of civil nuisance injunctions against prostitutes. This tactic would probably have also been found to infringe the jurisdiction of criminal law (indeed, when the Halifax council tried to use the same strategy to solve its prostitution problem the Supreme Court of Nova Scotia quashed the injunctions) but West End prostitutes in consultation with a local prostitutes' rights organization decided to move out of the area in order to avoid ongoing police harassment.

Some prostitutes moved to the area around Granville, Richards and Seymour Streets while others moved to the eastern part of Mount Pleasant, a working-class residential district outside the downtown area (Figure 12.2). It was not long before a group of Mount Pleasant residents cried foul and demanded action. In response the police relocated prostitutes in the north-western part of Mount Pleasant, an area occupied mainly by light manufacturing plants, but also containing a few residences. In so doing, they moved prostitutes to the doorstep of a woman who was to become one of the most vociferous campaigners against the street prostitution trade.

A new law making it an offence to 'communicate' in public for the purpose of buying or seling sexual services was enacted in December 1985. But despite vigorous enforcement of the new law it had only a short-lived impact on levels of street prostitution (Lowman 1989: 59–60). By the summer of 1986 Mount Pleasant activists were once again imploring government officials to do

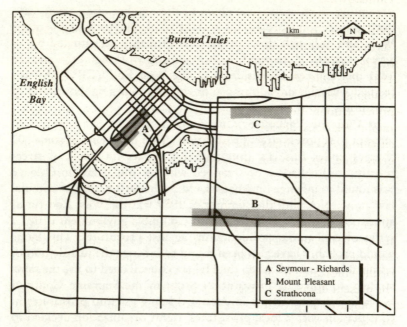

Figure 12.2 Vancouver prostitution strolls 1986–7

something to suppress street prostitution and began picketing street prostitutes in an attempt to drive them out of the area. In the summer of 1987 the police instituted a Prostitution Task Force (consisting of seven officers) in Mount Pleasant. Task Force activities included continual identity checks of prostitutes, charging them for any minor offences they might commit (jaywalking, littering, etc.) and parking marked patrol cars on the street corners where they worked. Men who appeared to be searching for a prostitute would be stopped and given all manner of traffic tickets. It did not take long for the number of prostitutes working in Mount Pleasant to decline. Task Forces were instituted again in 1988 and 1989 with the result that street prostitution in the area since 1987 has been minimal. But again the price paid for the success of this strategy has been the displacement of street prostitution to other areas. As prostitution has been suppressed in Mount Pleasant, so it has increased dramatically in Strathcona where the traditional area of street prostitution has expanded into a mainly residential area from its original location in the commercial area adjacent to Chinatown.

One can easily appreciate from the experience in Vancouver why local police would be far from amused by the idea that their 'policies "created" area differences in respect to prostitution'. Comparison of Figures 12.1 and 12.2 show that none of the areas used between 1981 and 1984 was still being used in 1986 and that much of the changing geography of street prostitution could be attributed to police actions. For the Vancouver police prostitution has been a public relations débâcle, and is perceived by them as a no-win situation. As one officer put it, 'It's like trying to trap mercury on a glass table top with the underside of a spoon'.

Police discretion and offender profiles

When the communicating law was enacted in 1985 the legislature for the first time made clear its intention to criminalize the public activities of customers as well as prostitutes. There are two aspects of the enforcement of the communicating law that are relevant to the present discussion: first, levels of enforcement against customers as compared to prostitutes, and second, the socio-economic status of the customers being charged.

Given that the various vagrancy laws which controlled street prostitution in Canada up to 1972 when the soliciting law was enacted explicitly applied only to women, and given that the soliciting law was applied only occasionally to customers,[10] there has been considerable interest in the manner in which the communicating law has been applied in this respect. In Vancouver in 1986 and 1987 of the 2,180 communicating charges laid, 532 involved customers as compared to 1,648 involving prostitutes – a ratio of approximately one to three. Law enforcement is proactive in the sense that it is carried out almost exclusively by police decoys. In the case of prostitutes, a plain-clothes policeman approaches a person he believes to be a prostitute and engages in a conversation in which he hopes to elicit the offer of sexual services for a fee. In the case of tricks a policewoman poses as a prostitute and waits for potential customers to approach her. There is substantial variation in the ratio of prostitutes to tricks charged in different police jurisdictions across Canada (Department of Justice Canada 1989), although in most (but not all) jurisdictions more prostitutes are charged than customers. While there are all sorts of arguments about what 'equal' enforcement might entail (cf. Lowman 1990) it should be obvious enough that the ratio of prostitutes to customers charged is almost entirely a product of the deployment of police decoys.

Not only does the deployment of police decoys influence the ratio of prostitutes to customers charged, but also it appears to have an important influence on the types of customers charged. Systematic head counts of prostitutes working in Vancouver's three main prostitution strolls in 1986 and 1987[11] as compared to the percentage of charges laid in each reveals an important discrepancy in the case of customers: very few of them were charged in the Richards-Seymour area, the stroll that contained the largest number of prostitutes (see Table 12.1).

Table 12.1 Area that charges were laid as compared to head counts of prostitutes (January 1986 to June 1987)

	Adult Pros		Youth Pros		Tricks		Counts
	No.	%	No.	%	No.	%	%
Mount Pleasant	94	46.6	20	37.7	154	66.9	38.75
Chinatown/E							
Hastings	33	16.3	7	13.2	64	27.8	12.5
Seymour/Richards	75	37.1	26	49.1	12	5.2	48.75
	202	100.0	53	100.0	230	100.0	100.00

Figure 12.3 shows the percentage of men charged with communicating residing in each of Vancouver's planning districts. If one were to draw a line down Cambie Street – which is, in a sense, Vancouver's class watershed and divides the city into two areas of roughly equal size – 91 per cent of the men charged with communicating in 1986 and 1987 came from the lower-income eastern half of the city. Figure 12.4 depicts the average family income in each district. Are we to assume that lower-income men have a greater propensity than upper-income men to purchase the sexual services of street prostitutes? To some extent the answer probably is yes. But some of this difference would also appear to be attributable to the geography of communicating law enforcement.

In Vancouver there are a variety of different classes of prostitutes, only some of whom work the streets. Probably at least as extensive as the street sex trade is that provided by escort agencies, of which

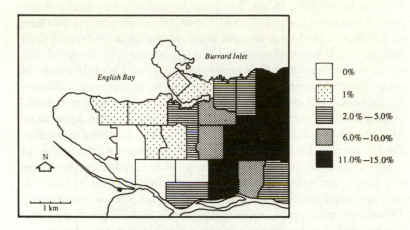

Figure 12.3   Vancouver men charged with communicating for the purpose
of purchasing sexual services

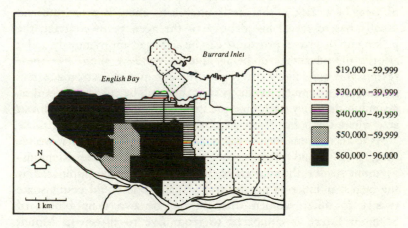

Figure 12.4   Average family income, Vancouver 1985

thirty-three were advertised in Vancouver's telephone Yellow Pages
in 1989. With an agency introduction fee plus the amount paid to
the escort, the minimum charge is usually in excess of $200. But as
well as this difference between street and off-street services there is

also a distinct price hierarchy in the three street prostitution areas. Prostitutes working in the Richards-Seymour area command the highest prices – usually $80 to $100 per 'trick'. Prostitutes working in Strathcona tend to command much lower prices – from $40 to $80 per 'trick' (depending on the type of service and the saleability of the women concerned). Women working the Richards-Seymour area – most of whom are quite young and wear expensive clothes – would balk at the idea of working in Strathcona, the skid row stroll. Mount Pleasant women formed an intermediate tier, but again some of them would not have been able to command anything more than the lowest fees. Besides the influence of the price of prostitutes on demand patterns, other factors may play a part in the street prostitute purchasing patterns that we see here. It may be that east side men were more likely to patronize Mount Pleasant prostitutes because the area is more accessible in terms of daily travel patterns; the Richards-Seymour area is more accessible to west side men.

While higher-income men presumably are much more likely than their lower-income counterparts to patronize escort service prostitutes, and thus not run the risk of being charged with communicating at all, there are still reasons to think that many of them patronize street prostitutes. The disadvantage of escort services from the point of view of a trick is that he is unable to chose a prostitute and usually has to reveal his identity to the agency. In contrast, the picking up of a street prostitute can be achieved anonymously and it affords the advantage of being able to window shop. For these reasons it is quite likely that higher-income men do patronize street prostitutes; it is much less likely that they will be charged, however, assuming that they patronize prostitutes in the Richards-Seymour area where the police rarely deploy female decoys.

As to the reasons for the pattern of law enforcement that we see here, they would seem to be quite mundane. The Richards-Seymour stroll is the smallest of the three areas and is populated by the largest number of prostitutes. Usually one could count some twenty to thirty women working on the two main blocks of Seymour Street as compared to from five to fifteen in Mount Pleasant and Strathcona. There are only sixty female police-officers in the Vancouver Police Department, of whom relatively few are young enough to pose successfully as decoys (especially in the Richards-Seymour area). Police reason that if they place a policewoman in this area it takes hours for her to be approached by a trick, whereas in Mount Pleasant and Strathcona it did not take very long at all. After the first few months of communicating law enforcement police say they gave up enforcing the law against tricks in the Richards-Seymour area. The result would seem to be that lower-income men are much more likely than their higher-income counterparts to be charged with communicating in public for the purpose of purchasing sexual services.

CONCLUSIONS

Again it should be emphasized that the example presented here is not meant to be generalized to crimes reported to the police by the public, although one may well see similar patterns when it comes to drug offences (cf. Boyd *et al.* 1987: 490–1) and other targetless crimes. But even when it comes to what is commonly referred to as reactive policing, there are certain senses in which police practices can be seen to influence crime rates.

In examining the impact of 'methodical suspicion' and 'incongruity procedures' on crime rates Mawby's work has helped to demonstrate that *the role of the public is down played in work which emphasizes the role of the police in identifying criminals and crimes.* But cross-sectional analyses designed to make this point are unable to test various kinds of *amplification of deviance* hypotheses in which incongruity procedures and methodical suspicion may play an important part in forging criminal identities even if they do not 'produce' crime in a narrow empirical sense. But even when it comes to the role of the police in identifying crimes and criminals it would be premature, especially in the case of proactive police work of the sort described above, to rule out the influence of police behaviour on patterns of crime and patterns of offender residence. On top of all these considerations are broader issues – such as the impact of police practices on public definitions of crime and the differential allocation of resources to the policing of different kinds of crimes – that remain largely beyond the reach of the type of analysis provided by Mawby, but which are nevertheless germane to understanding 'crime' rates.

Given the debate that still simmers in criminology about the interpretation of crime statistics, a degree of caution would seem to be warranted. Perhaps Bottomley and Coleman are closest to the mark when they suggest that while police generated crime statistics are not simply an epiphenomenum of police work, the process of record keeping is indispensable to their understanding (Bottomley and Coleman 1981: 146). Any undestanding of this process will require both institutional and social-structural levels of analysis, both of which are still in their infancy.

NOTES

1 In Matza's (1969) hypothetical account methodical suspicion arises from being 'known to the police', known in the sense that a particular identity or resemblance – involving such factors as gender, age, race and dress – produces suspicion. Incidental suspicion, in contrast, arises from the circumstances linking a suspect to a crime; the person holding the smoking gun over the corpse, for example, is suspected of firing the deadly bullet.

2 That type of police work where police themselves identify suspects or discover crimes as opposed to reacting to citizen crime reports or information about the identity of suspects.

3 Offences involving illicit drugs, prostitution and other 'vices' have been referred to as 'victimless' crimes because there is no specific victim to report an incident to the police. But it seems reasonable to argue that there are generalized victims of such offences – for example, people living in areas where street prostitution occurs – or that there can be a sense in which the offenders themselves are victims. The distinctive feature of such offences is that they do not have a specific *target* in the sense that most property and violent offences do.

4 In Canada in 1986, for example, there was a total of 2,277,749 *Criminal Code* offences. These included 365,140 break and enters, 85,585 motor vehicle thefts, 23,268 robberies and 569 homicides. Theft (773,257 offences) and wilful damage (327,644 offences) accounted for nearly half the total.

5 Mawby (1979: 30–1) claims that such processes are important only in the case of extremely notorious estates. But such practices may be more important than he suggests when it comes to the operation of 'sus' laws and other all-purpose control devices widely employed by patrol police.

6 In Chicago Reiss (1971: 96) estimated that 1 per cent of the time spent in preventive patrol was occupied in this way, while McCabe and Sutcliffe (1978: 9) suggest that only 6 per cent of police time is spent on incidents that end up being defined as criminal or a breach of public order.

7 The finding is tentative because Westwood notes that it is possible that geographic variations in crime seriousness explain the pattern of cautioning that he describes.

8 Although one of the best indicators of crime reporting is the perceived 'seriousness' of the offence (cf. Gove *et al.* 1985), the fact remains that most crime is not serious according to the particular vision of 'seriousness' employed by the authors who make this point. Also at issue here is how the notion of seriousness is socially constructed.

9 This law made it an offence to 'solicit anyone in a public place for the purpose of prostitution'. What had been at issue in the courts since the law's enactment in 1972 was the exact meaning of the term 'solicit'. In

1978 the Supreme Court of Canada ruled that it had to be established that a prostitute was 'pressing and persistent' in her (or his) conduct to be found guilty of 'soliciting'. This meant that police could no longer secure a conviction simply by establishing that a prostitute had offered to provide sexual services for a price. Vancouver police concluded that this decision rendered the soliciting law unenforceable (Fraser Committee 1985: 419–29).

10 In Ontario the courts ruled that the soliciting law could be applied to customers (although only a handful of charges were actually laid) while in British Columbia the courts ruled that it could not (Fraser Committee 1985: 420).

11 The head counts involve a tally of the number of visible prostitutes on a single traverse of the streets in each stroll at various times through the day and week. Long-term trends are ascertained from counts taken each Thursday night between 10.00 and 11.00 p.m. (these are available from November 1982 to the present).

REFERENCES

Biderman, A.D. and Reiss, A.J. (1967) 'On exploring the dark figure of crime', *Annals of the American Academy of Political Science* 374: 1–15.

Bottomley, A.K. and Coleman, C.A. (1981) *Understanding Crime Rates: Police and Public Roles in the Production of Official Statistics*, Westmead, Hants: Saxon House.

Bottomley, A.K. and Pease, K. (1986) *Crime and Punishment: Interpreting the Data*, Milton Keynes: Open University Press.

Bottoms, A.E. and Xanthos, P. (1981) 'Housing policy and crime in the British public sector', in P.J. Brantingham and P.L. Brantingham (eds) *Environmental Criminology*, Beverly Hills, Calif.: Sage.

Boyd, N., Lowman, J. and Mosher, C. (1987) 'Case law and drug convictions: testing the rhetoric of equality rights', *Criminal Law Quarterly* 27: 487–511.

Brantingham, P.J. and Brantingham, P.L. (1981) 'Introduction: the dimensions of crime', in P.J. Brantingham and P.L. Brantingham (eds) *Environmental Criminology*, Beverly Hills, Calif.: Sage.

Cicourel, A.V. (1968) *The Social Organisation of Juvenile Justice*, New York: Wiley.

Department of Justice Canada (1989) *Street Prostitution, Assessing the Impact of Law: Synthesis Report*, Ottawa: Department of Justice Canada.

Ditton, J. (1979) *Contrology: Beyond the New Criminology*, London: Macmillan.

Ericson, R.V. (1982) *Reproducing Order: A Study of Police Patrol Work*, Toronto: University of Toronto Press.

Farrington, D.P. and Dowds, E.A. (1985) 'Disentangling criminal behaviour and police reaction', in D.P. Farrington and J. Gunn (eds) *Reactions to Crime: The Police, Courts and Prisons*, Chichester: Wiley.

Fraser Committee (1985) *Pornography and Prostitution in Canada*, Ottawa: Minister of Supply and Services Canada.

Gill, O. (1977) *Luke Street*, London: Macmillan.

Gove, W.R., Hughes, M. and Geerken, M. (1985) 'Are uniform crime reports a valid indicator of index crimes? An affirmative answer with minor qualifications', *Criminology* 23 (3): 451–501.

Jones, T., MacLean, B. and Young, J. (1986) *The Islington Crime Survey*, Aldershot: Gower.

Kinsey, R., Lea, J. and Young, J. (1986) *Losing the Fight against Crime*, Oxford: Blackwell.

Kitsuse, J.I. and Cicourel, A.V. (1963) 'A note on the use of official statistics', *Social Problems* 11: 131–9.

Lea, T. and Young, J. (1984) *What is to be Done about Law and Order?*, Harmondsworth: Penguin.

Lemert, E.M. (1951) *Social Pathology*, New York: McGraw Hill.

Lowman, J. (1982) 'Crime, criminal justice policy and the urban environment', in D.T. Herbert and R.J. Johnston (eds) *Geography and the Urban Environment*, vol. 5, Chichester: Wiley.

—— (1989) *Street Prostitution, Assessing the Impact of the Law: Vancouver*, Ottawa: Department of Justice Canada.

—— (1990) 'Notions of formal equality before the law: the experience of street prostitutes and their customers', *Journal of Human Justice* 1 (2): 55–73.

McCabe, S. and Sutcliffe, F. (1978) *Defining Crime: A Study of Police Decisions*, Oxford: Blackwell.

Maclean, B. (forthcoming) 'In partial defence of left realism: some theoretical and methodological concerns of the local crime survey', *Contemporary Crises*.

Matza, D. (1969) *Becoming Deviant*, Englewood Cliffs, NJ: Prentice Hall.

Mawby, R. (1978) 'A note on domestic disputes', *Howard Journal* 17: 160–8.

—— (1979) *Policing the City*, Aldershot: Gower.

—— (1989) 'Policing and the criminal area', in D.J. Evans and D.T. Herbert (eds) *The Geography of Crime*, London: Routledge.

O'Brien, R.M. (1985) *Crime and Victimization Data*, New York: Sage.

Pepinsky, H.E. (1987) 'Explaining police reported crime trends in Sheffield', *Contemporary Crises* 11 (1): 59–73.

Reiss, A.J. (1971) *The Police and the Public*, New Haven, Conn.: Yale University Press.

Sacks, H. (1972) 'Notes on police assessment of moral character', in D. Sudnow (ed.) *Studies in Social Interaction*, New York: Free Press.

Sparks, R.F., Genn, H.G. and Dodd, D.J. (1977) *Surveying Victims*, London: Wiley.

US Bureau of Justice Statistics (1988) *Bulletin*, Washington, DC: US Department of Justice.

Westwood, D. (1990) 'Adult cautioning', *Policing* 6 (1): 383–98.

Wilkins, L.T. (1964) *Social Deviance*, London: Tavistock.

Willis, C. (1983) *The Use, Effectiveness and Impact of Police Stop and Search Powers*, Research and Planning Unit Paper 15, London: Home Office.

Young, J. (1987) 'The tasks facing a realist criminology', *Contemporary Crises* 11 (4): 337–56.

Part III

CRIME AND POLICING: POLICY PERSPECTIVES

13

CHANGING PERSPECTIVES ON CRIME PREVENTION
The role of information and structure

Kevin Heal

MARKETING AN IDEA

The 1980s witnessed a number of significant shifts in society's response to crime. Those seeking explanation of the change must put the clock back to the early 1980s, and the beginning of Lord Whitelaw's period of office as Home Secretary. Allowing for some oversimplification, the most significant development of the time was the agreement secured across Whitehall that crime reduction was not solely the responsibility of the Home Office, but a concern of all central government departments (Home Office 1984). At local level, so the argument ran, all agencies had a part to play, not only the police. Criminologically the move made good sense reflecting as it did the complex aetiology of crime, and the limitations of society's traditional response to it.

Several factors underlay the new thinking. While it is not possible to separate the predisposing from the precipitating, it is almost certainly true that increasing pressure on the criminal justice system encouraged the search for a broader response to crime.

The pressure on the machinery of justice had many forms. The rise in recorded crime experienced in the early 1980s was no new phenomenon: crime had drifted steadily upwards at a rate of 5–7 per cent a year since the 1950s. However, convincing arguments suggested that while a proportion of the increase in crime may have been illusory, stemming from changes in reporting and recording procedures – a point made by Rutter and Giller (1984) in respect of juvenile crime – crime itself was on the increase. Moreover, social surveys suggested that the recorded crime figure in the early 1980s (a little under 3 million offences per year) was a substantial underestimate of the problem. About half of all burglaries it was

257

suggested went unrecorded and perhaps as many as three-quarters of woundings and sexual attacks (Home Office 1984).

Increasing demand on the criminal justice system sharpened scrutiny of the system itself. The effectiveness of policing, sentencing and imprisonment were all found wanting with a reforming zeal which perhaps led to a degree of oversimplification (Folkard *et al.* 1976; Brody 1976; Morris and Heal 1981). None the less, the basic premise of these iconoclastic studies found favour with an interdepartmental group of senior civil servants convened to take forward Lord Whitelaw's initiative. Indeed the annex to the report of that group notes that 'the scope for reducing crime through changes to the criminal justice system itself is very limited' (Home Office 1984).

Faced with a growing demand upon the system, which at best appeared to have no capacity for greater impact, the interdepartmental group commended to the readers of its report a preventive response to crime which focused on the reduction of opportunities to commit crimes.

The report of the interdepartmental group was received with a degree of enthusiasm. Indeed, it would have been surprising had it been otherwise. Opportunity reduction, soon to be marketed as 'situational prevention', had a common-sense credibility for the practitioner. Research findings lent authority to an old idea; its message was simple, and its promised impact immediate and demonstrable in terms of local reductions in crime figures. Moreover, it served to fill the vacuum created by the demise of the treatment model of crime control, and to lighten the gloom of the 'nothing works' pessimism of the late 1970s.

Patterns of crime were discovered, and their spatial distribution explained in terms of the opportunities created by environmental factors: areas of high car crime were associated with insecure car parks and the prevalence of unlockable or unlocked cars; high burglary rates were associated with insecure pre-payment fuel meters, or weak door or window frames (Hill 1986; Clarke and Mayhew 1980).

As far as criminological theory was concerned a niche for situational prevention was quickly found within contemporary perspectives of behaviour. Links were made with rational choice theory in the psychological literature of the 1950s; with the sociology of the late 1960s and with the economics of the 1980s. As Bennett (1986) showed these provided a theoretical framework for the situational model developed by Clarke and Cornish (1985). Bennett, drawing a distinction between the initial decision to offend and the final decision to offend, argued that, since it was unlikely the existence of criminal opportunities would motivate the un-motivated, situation prevention should be directed at the final decision. Here it was possible to alter the balance of cost and rewards by manipulating the physical situation.

Against this background the crime prevention bandwagon, engineered from notions of opportunity reduction; from design solutions to crime; physical security; formal and informal surveil-lance and the removal of targets, rolled with some speed.

SECOND THOUGHTS

The activity associated with situational prevention brought, inevit-ably perhaps, criticism which focused upon the shortcomings of the approach.

For some the preventive response was seen as a policy of despair, although this argument was rebutted by Laycock and Pease (1985) in an article on the probation service and crime prevention. Others accepted the value of prevention but questioned the emphasis then being placed on opportunity reduction. Rutter and Giller (1984), for example, writing on the prevention of juvenile crime around the time of the interdepartmental group's report, commented on the importance of improved school ethos and peer influences, and the declaratory role of teachers and parents in expressing firm and unequivocal disapproval of delinquency. Some pointed to the contribution of pre-school education to sound social adjustment, while a study by Wilson (1980) identified the problems created by a lack of proper parental supervision. The National Association for the Care and Resettlement of Offenders (NACRO), drawing on the report of the Cunningham Road project (NACRO 1980), empha-sized the importance of addressing crime from the perspective of the community.

These views which, had they been drawn together could have provided an alternative approach to prevention, in fact did little or nothing to check the growing belief in opportunity reduction as the viable response to crime. They did, however, raise some pertinent questions.

Too little crime?

For many the flaw running through rational choice theory (one interpretation of situational prevention) was its failure to offer a sufficient explanation of criminal behaviour. Some, indeed most people, do not steal cars from unlit streets, or break into insecure premises despite the millions of opportunities that exist. Environmental determinism (that is the influence of environmental opportunities on crime) may be a necessary condition for crime to occur but is not of itself a sufficient condition. Tuck, writing in 1987, reflected the views of many (Home Office 1987). There was substance in the criticism. There is clearly a difference between those who see an open window as a passport to crime, and those who view the same window as an aid to ventilation. It is, it can be argued, the individual's learned perception of a situation which is of critical importance not the situation itself, and the readiness of the individual to turn to crime to meet specific needs.

Limits of pragmatism

For those supporting situational prevention the weighting to be attached to the various elements in the criminal equation (opportunity, learned perception and propensity to commit crime) was determined, not by theoretical argument, but by pragmatism. The individual's moral code, previous learning experience and psychological and family background were, it was recognized, less amenable to intervention and change than the opportunity he or she was confronted with. Understandably, therefore, in a climate (described earlier) where practical impact was required, criminal opportunities and situational variables attracted particular attention. The supporters of situational prevention, while not claiming the approach explained all crime argued, with justification, that it supported a far wider range of practical preventive measures than any competing approach. Support for this view came first from the work of Clarke, as Head of the Home Office Research and Planning Unit (Clarke and Mayhew 1980), and subsequently from the newly formed Home Office Crime Prevention Unit. By 1986 Trasler felt able to conclude that, in terms of practical impact, the thinking underlying situational prevention was ahead of most other theories.

Yet, as Trasler (1986) also noted, situational prevention (at least in its present form) was of *limited* effect when dealing with many expressive crimes: most homicides, rapes and domestic violence. Clarke and Cornish (1986) challenged this, and it is certainly true that opportunity reduction measures effectively reduce violence where the likely location is known and open to intervention (for example in banks, building societies and post offices). Where the economic gains from violent crime are exceptionally high, however, or where the readiness to commit crime is fuelled by an intense and enduring social conflict (domestic violence), or where the location within which crime occurs is not amendable to significant environmental change (much street crime) the potential impact of situational prevention seemed limited.

Displaced crime?

For the most part the issue of displacement (one of the most telling criticisms made of situational prevention) continues to be debated not on the basis of data or careful argument, but from the position of personal hunches, and predilections about the origins of human behaviour. Thus, if wedded to the drive theories of the 1940s displacement becomes a powerful argument against opportunity reduction as a form of crime control. Sympathy with this approach underpins the extreme case pessimism referred to by Clarke and Cornish (1986); it argues for a form of prevention which is not concerned with blocking immediate behaviour but with influencing the hidden roots of criminality. Conversely those who attach greater importance to the situational variables in the behaviour equation, while not discounting of displacement, do not see it as undermining their approach to prevention. Thus, it can be argued, that blocking an opportunity for crime may well be sufficient to stop the offender comitting crime at all: crime abatement. Within this model the opportunity for crime is the major determinant of behaviour open to influence and the drives, inclinations, and perceptions leading to the exploitation of that opportunity of lesser importance.

Research findings being contradictory do little to resolve the argument. The clutch of studies demonstrating that preventive interventions displace crime can be easily balanced by research to the contrary. The absence of consensus suggests that the issue is complex, and the current analysis of it oversimple. More attention probably needs to be given to the characteristics of the criminal, and the opportunities that exist for alternative forms of behaviour to be pursued.

The uncertainty surrounding the displacement argument is understandable for while it is comparatively straightforward to identify situations where displacement will occur and to demonstrate that it has taken place, it is far more difficult to prove that preventive interventions can take place without displacement occurring. The network of social interaction is so complex that the precise consequences of blocking an opportunity through preventive measures can never be monitored comprehensively. If an opportunity is blocked the potential criminal will do something for good or ill. At best the chosen form of behaviour may be legitimate and crime abatement or desistence may occur. Clearly this can be seen as a form of displacement, however the term tends to be used when the alternative form of behaviour is criminal. Within this category it is possible to identify different forms of displacement (Hakim and Rengert 1981). However, these refinements do not cope with the problem that consequences of intervention (opportunity reduction) could easily appear well beyond the researcher's field of study. This means that it is impossible to refute completely the argument that situation measures merely deflect crime.

Despite differences of view regarding the displacement of crime most would agree that at the heavy end of criminal behaviour it is almost certainly the case that some activities are so lucrative that offender displacement is almost certainly inevitable following intervention. Barr and Pease (1990) illustrate the point by reference to drug trafficking where the rewards are so high, and the pool of potential offenders so great, that there will be no shortage of individuals willing to commit a crime regardless of what happened to the predecessors, or the situations within which they operated.

Discussions of displacement also lead to a degree of agreement regarding the comparative flexibility of human behaviour. Thus where an opportunity to commit crime is barred, offenders will either desist, that is will return to non-criminal behaviour, or shift their behaviour to some other form of criminal activity, or some other target. What they do not do, or rarely so, is to continue on the same behavioural course. This suggests there may be considerable scope for managing the crime problem in a way that other aspects of social behaviour are managed.

Political bias

King (1989), reflecting on crime prevention activity in the UK over the past ten years, questions the soundness of the preventive measures he observed from a different perspective. In particular he challenges their principal characteristics which he sees as reflecting those of the Tory administration which governed the country during the period. King draws to the attention of his reader the individualism, consumerism and self-protection which he sees as underpinning the crime prevention of the 1980s. This is at odds, King suggests, with protecting the weak, and the idea of working with communities and their democratically elected representatives (local government). King's arguments contain within them some insights but tend to drown the crime prevention baby in the political bathwater. He ignores, for example, the fact that neighbourhood watch, initially based on locks, bolts and bars and property marking, is frequently transformed into an activity concerned with community social cohesion and informal support. Similarly the assertion that the Safer Cities Programme (discussed later) is a deliberate attempt by central government to bypass local democratic structures, fails to reflect the principles and practice of the programme itself. Moreover, for reasons that are not entirely clear, King chooses to ignore the extensive research literature illustrating the effectiveness of opportunity reduction as a mechanism of crime prevention, and to advocate in its place an approach for which there is little or no evidence of effect.

A PARTIAL ANSWER

The crime prevention of the early 1980s prompted a series of questions and criticisms ranging from those driven by commonsense to others inspired by political dogma. If nothing else, the criticisms demonstrated that situational prevention, the mainspring of the early 1980s activity, was only a partial solution to the task of preventing crime. This recognition fuelled alternative thinking which focused not upon the physical situation within which crimes occur or the victims of those crimes, but upon the perpetrators of the crimes, and the social policies which bear on the lives of the perpetrators and the communities within which they live.

This was not a new idea. The Victorian answer to crime lay in the notion of the criminal personality which resulted in the individual habitually breaking the law. Criminals needed to be taught to be morally better, hence the emphasis on chaplains, prison chapels and moral education. While the language has changed, dissatisfaction with situational prevention again focused attention on the offender, but on this occasion on the need to find ways of strengthening the individual's ties with the community, so allowing him or her to engage with the larger social groupings of family kin, school, stable neighbourhood and work-place. This has been coupled with an interest in the community itself, and the view that communities are capable of exerting social control through informal mechanisms.

Interest in strengthening informal social control by the community, and by drawing offenders into community networks, is not far removed from the view that crime should be tackled through improved parenting, intensive education and counselling for troublesome children in an attempt to cope with the adverse social, economic and psychological factors associated with the early childhood of offenders (Farrington 1989). These and related thoughts have been drawn together under the broad heading of social crime prevention. Unfortunately while they are strong ideologically, and in some respects theoretically, limited progress has been made in implementing them in disadvantaged areas. What is missing is the empirical evidence that investment in them will reduce crime (Bottoms 1989).

Failure to pull these thoughts together has been costly. Their advocates have pointed to the limitations of situational prevention and to the gaps in that approach to crime. In so doing they have almost certainly undervalued the contribution situational prevention can make; more importantly they have failed to provide an alternative or complementary programme. This is clearest in respect of neighbourhood watch schemes which, having locked and bolted and begun to engage in community activities, still await a social crime prevention agenda.

For a time, therefore, it appeared that crime prevention may be in danger of stalling. A belief in the importance of social cohesion (as an aid to crime reduction), for example, started to find expression in the expectation that somehow (in ways unspecified) communities would 'pull themselves together' and extend effective networks and generate anti-crime attitudes in young people. Currie (1988) refers to this as Phase One crime prevention and characterizes the thinking thus: 'if only people would get their act together – everything would be alright'. He refers to the benign cycle of prevention whereby attitudes improve, behaviour improves, and individuals identify with the community. Unfortunately, it always remained far from clear how individuals living in the hardest pressed parts of the country (37 per cent of burglaries and 33 per cent of theft from the person and robberies occur in run-down inner city areas although such areas only contain 12 per cent of the country's households) could hope to 'get their act together' and so reduce crime (Hope and Hough 1988). Arguably the hallmark of Phase One prevention was an abdication of responsibility by statutory and voluntary agencies alike.

A REVISED AGENDA

The close of the 1980s saw the emergence of a more realistic approach resting on the belief that the development of crime prevention could benefit from the creative tension of schism rather than suffer from its excesses. What began to emerge was the hunt for a strategy which recognized the importance of both precipitating factors (e.g. the criminal opportunities for crime) and predisposing influences (the social and psychological needs and learned perceptions and attitudes of the individual which encourages him to exploit those opportunities). Within this broader strategy prevention might be set alongside the processes of detection, investigation and court disposal on the one hand, and care for the victim on the other, so forming a composite strategy for crime.

The synthesis of measures, and the development of a more comprehensive approach to crime, started to emerge towards the end of the decade not as a result of a theoretical or a research breakthrough, but from a simple but significant change in the structure for delivering crime prevention. The key elements of the new schemes starting to appear at this time were:

the preparation of a crime profile for an area on the basis of which a strategic crime prevention plan was devised; and placing responsibility for the plan and its implementation in the hands of a steering committee which brought together different agencies and professional groups each having their own particular view of prevention.

Taken together these characteristics (which were to be found in the Home Office Five Towns Initiative: Home Office 1988) forced schemes to offer a composite response to an area crime problem. Since the steering committees were made up of individuals with different skills and perceptions the plans produced of necessity avoided the artificial distinction between physical prevention and social measures. Physical security measures, bringing instant impact, and so securing credibility for the projects, provided a foothold on which those involved in the Five Towns Initiative (and similar schemes) could build projects aimed at changing perceptions, creating skills, opening opportunities to legitimize behaviour which inevitably proceeded in slower time. Members of hard-pressed communities were no longer expected to 'bootstrap' their way out of crime problems. Rather communities were examined in terms of the facilities they provided for employment, leisure and, through the school, church and family the perceptions and attitudes that were disseminated to their young people.

This change in structure of local prevention was taken forward by the more ambitious Home Office Safer Cities Programme where again the area focus, crime profile and multi-agency steering group encouraged the integration of a range of measures to deal with criminal activities. Within this structure, therefore, the earlier schism between physical and social measures against crime was reduced (Home Office 1989).

In important respects, therefore, crime prevention may well have recovered from the difficult ground it began to enter in the mid-1980s and to have found a more profitable route, or at least to be in sight of it. Whether or not it advances to a more sophisticated and more effective form of activity depends almost entirely upon whether a number of major obstacles are overcome in the 1990s. This chapter concludes by reviewing some of the obstacles.

OBSTACLES TO PROGRESS

The start of the 1990s sees a surprising degree of unanimity about crime prevention: academics, policy-makers and practitioners alike agree on the need

1 to integrate social and physical measures against crime in 'a careful and creative synthesis of the two approaches' (Bottoms 1989)
2 to establish crime prevention in high crime areas
3 to move from a project-based method of delivering crime prevention to a policy/process approach taken forward through reliable and enduring local structures, by able leadership, and with adequate resources.

The obstacles to the development of prevention are many and varied but among the more important are the lack of reliable information about crime, and the lack of local structures to support the delivery of prevention.

Lack of information

The success of most endeavours depends to a large extent upon the quality and the quantity of the information upon which they are based. The information stock on any subject sets the parameters of thinking and activity, indeed both are imprisoned within the bounds of information. Hence the value of speculative theory and the tendency (appropriate in this case) for the media to refer to a 'breakthrough' when years of patient research increase the breadth of understanding.

Those concerned with the control of crime have been, and to some extent still are, seriously hampered both by the paucity of information on crime, criminals and victims and the failure to establish integrated data sets on the basis of the material that is available. Criminologists, practitioners and policy-makers have laboured valiantly with official statistics, a source of data more informative of the criminal justice system than the characteristics of crime. The National Crime Surveys of the 1980s broke the mould and offered new perspectives, as did the few local surveys which followed in their wake. Yet despite these efforts to foster in-depth analysis of crime data prevention remains starved of reliable information.

The consequences of this lack of information are far-reaching, and some have already been touched on in this chapter. It was suggested earlier, for example, that it was the absence of sound information which allowed the displacement argument to be driven by armchair theorizing and personal hunches about human behaviour. Only recently has work by Barr and Pease (1990) taken the argument forward by drawing together crime information from a variety of sources. But unfortunately even this has already been misrepresented in the press (Hyder 1989) simply, as far as one can tell, for the sake of scoring journalistic points: an exercise which illustrates the damage misinformation can do to crime prevention.

Further examples of the costs of working from an inadequate information base can be readily found. Bennett (1990), writing of his evaluation of selected London neighbourhood watch schemes, argued that since the schemes studied were devised and implemented with little or no understanding of the crime they were supposed to address, or the capability of the community to control it, their failure was unremarkable.

Preventive activities for young people further illustrate the point: a failure to understand the would-be offender's perception of crime and its rewards can easily result in the flourishing of inappropriate strategies. That many crimes involve the offender in highly dangerous situations should tell us something about the rewards, excitement and status the successful crime brings. We are told, for example, of young people repeatedly putting their lives at risk to spray graffiti on transportation systems; or – in Northern Ireland – persisting in taking and driving away cars (joy riding) when the local vigilante sanction against this offence is known to be knee-capping. If these are the risks that offenders are willing to take to secure criminal rewards it seems at least likely that anti-crime diversionary activities will have to offer more than coffee and a pool table.

These examples illustrate the importance of systematically collecting reliable information if effective preventive measures are to be developed.

The imposition of crime pattern analysis on recorded crime statistics marked a notable step forward in filling the crime information gap. The work of Engstad and Evans (1980) set the scene giving material both for the practitioner and theorist to draw on. The hallmark of the Five Towns Initiative and the Safer Cities Programme is the dependence of both on the area crime profile, yet despite this promising beginning, and the emergence of relevant information technology, progress has been remarkably slow, and for the most part spatial distributions of crime remain based on samples of limited size, with few efforts being made to check the reliability or stability of the patterns observed. Attempts to match crime patterns with patterns in socio-demographic data have also been limited.

The lack of data, together with a failure to integrate and fully analyse the material that is available, places limits on the extent to which crime prevention theory and practice can develop in the coming decade.

In the absence of structure

Most would also agree that one of the key obstacles to the development of prevention is the absence of local structures for delivering crime prevention as a service. Six years after the issue of the joint circular on crime prevention by central government local activity remains fragmented and patchy, duplication of effort is not uncommon, and co-ordination limited in its scale. Evidence of this, if evidence is needed, came from the government's own need to establish *ad hoc* structural arrangements for the Five Towns Initiative and, subsequently, the larger Safer Cities Programme. The absence of structure, with the associated leadership, funding and information problems, means that it is not unusual for crime prevention initiatives to be peripheral, fading when key staff move on, or when initial funding ceases.

The lack of structural arrangements reflects a lack of agreement as to which professions are responsible for the co-ordination of preventive activities. Again the lack of information about crime hinders the debate. Where crime prevention is seen primarily as locks, bolts and bars, the traditional activities and expertise of the police place them in the lead. However, as thinking broadens and the relevance of other agencies to the prevention of crime becomes apparent, the leadership question becomes less readily answered. Given the nature of crime and the complexity of its origins, it is likely that the only satisfactory solution will be a co-operative one which brings the skills of different organizations and individuals together. This suggests that it probably does not matter which agency leads provided those concerned have the skills to bring different professions and bodies together in an effective way. This calls for welding together existing structures (police, local authority and probation) within an effective forum for planning and action at county, district/borough and street level. The Home Office revised guidance on prevention provides a basis from which this complex question might be addressed (Home Office 1990).

In summary there is a clear agenda for prevention but action in pursuit of it is likely to be limited in at least two important ways. First, it could be trapped within existing information and analysis systems with the result that personal preference, anecdote and 'common-sense' guide many decisions which would be better based on reliable data. Next, assuming information is released and a balanced form of prevention incorporating both physical and social elements developed, prevention could falter unless local structures are put in place to support it.

This review of the ebb and flow of crime prevention thinking over the past decade leads to a position where information and structures can be seen as central to the continued development of preventive action against crime: information to guard against the type of careless thinking and personal prejudice which can so easily lead prevention into a cul-de-sac, and structures which are capable of developing the ideas flowing from the information. In keeping with this thinking the recently issued Home Office Circular on crime prevention and accompanying booklet (Home Office 1990) identifies structure and information as among the key elements experience has shown as necessary to any successful crime prevention project, programme or policy. Drawing on examples of successful local schemes the publication illustrates how some agencies have found ways of overcoming the obstacles that may easily hinder the development of prevention.

REFERENCES

Barr, R. and Pease, K. (1990) 'Crime displacement and placement', in M. Tonry and N. Morris (eds) *Crime and Justice*, vol. 12, Chicago: University of Chicago Press.

Bennett, T. (1986) 'Situational crime prevention from the offenders perspective', in K. Heal and G. Laycock (eds) *Situational Crime Prevention from Theory into Practice*, London: HMSO.

—— (1990) *Evaluating Neighbourhood Watch*, Aldershot: Gower.

Bottoms, A. (1989) *Crime Prevention Facing the 1990s*, Glasgow: James Smart Lecture.

Brody, S.R. (1976) *The Effectiveness of Sentencing*, Home Office Research Study 35, London: HMSO.

Clarke, R.V.G. and Cornish, D.B. (1985) 'Modeling offenders decision: a framework for policy and research', in M. Tonry and N. Morris (eds) *Crime and Justice: An Annual Review of Research*, vol. 6, Chicago: University of Chicago Press.

—— (1986) 'Situational prevention, displacement of crime and rational choice theory', in K. Heal and G. Laycock (eds) *Situational Crime Prevention from Theory into Practice*, London: HMSO.

Clarke, R.V.G. and Mayhew, P. (eds) (1980) *Designing Out Crime*, London: HMSO.

Currie, E. (1988) 'Two visions of community crime prevention', in T. Hope and M. Shaw (eds) *Communities and Crime Reduction*, London: HMSO.

Engstad, P. and Evans, J. (1980) 'Responsibility, competence and police effectiveness', in R.V.G. Clarke, and J.M. Hough (eds) *The Effectiveness of Policing*, Aldershot: Gower.

Farrington, D.P. (1989) *Cambridge Study in Delinquency Development: Longterm Follow-Up*, Cambridge: Institute of Criminology.

Folkard, M.S., Smith, D.E. and Smith, D.D. (1976) *Intensive Matched Probation and After-Care Treatment*, vol. II, Home Office Research Study 36, London: HMSO.

Hakim, S. and Rengert, G.F. (1981) *Crime Spillover*, Beverly Hills, Calif.: Sage.

Hill, N. (1986) *Prepayment Coin Meters: A Target for Burglary*, Crime Prevention Unit Paper 6, London: HMSO.

Home Office (1984) *Report of an Interdepartmental Group on Crime*, London: HMSO.

—— (1987) *Crime, Criminality and Crime Prevention*, London, unpublished discussion paper.

—— (1988) *The Five Towns Initiative*, London: HMSO.

—— (1989) *Safer Cities, Progress Report 1988–89*, London: HMSO.

—— (1990) *Partnership in Crime Prevention*, London: HMSO.

Hope, T. and Hough, M. (1988) 'Area, crime and incivility: a profile from the British Crime Survey', in T. Hope and M. Shaw (eds) *Communities and Crime Reduction*, London: HMSO.

Hyder, K. (1989) 'Sharing the crime load', *Police Review* November: 2,274–5.

King, M. (1989) 'Social crime prevention à la Thatcher', *Howard Journal* 28: 291–312.

Morris, P. and Heal, K. (1981) *Crime Control and the Police*, Home Office Research Study 67, London: HMSO.

National Association for the Care and Resettlement of Offenders (1980) *Cunningham Road Project*, London: NACRO.

Laycock, G. and Pease, K. (1985) 'Crime prevention within the Probation Service', *Probation Journal* 32: 43–7.

Rutter, J. and Giller, H. (1984) *Juvenile Delinquency: Trends and Perspectives*, Harmondsworth: Penguin.

Trasler, G. (1986) 'Situational crime control and rational choice: a critique', in K. Heal and G. Laycock (eds) *Situational Crime Prevention from Theory into Practice*, London: HMSO.

Wilson, H. (1980) 'Parental supervision: a neglected aspect of delinquency', *British Journal of Criminology* 20: 20–35.

14

THEMES AND VARIATIONS IN NEIGHBOURHOOD WATCH

Trevor Bennett

In terms of public enthusiasm and support neighbourhood watch (NW) must be one of the most successful crime prevention programmes ever created.[1] The first NW scheme in Britain was launched in 1983 (less than seven years ago at time of writing); it is now estimated that there are over 74,000 NW schemes in the country (*Observer* 1989). The results of the second British Crime Survey (BCS) showed that 62 per cent of residents would be prepared to join a NW scheme if one were in their area (Hough and Mayhew 1985) and the results of the third survey showed that 90 per cent of the population of England and Wales had heard of NW. These figures are impressive by any standards and indicate a strong willingness among the public to work alongside their local police in the prevention of crime.

There is less optimism, however, about the success of these schemes in terms of their level of implementation or in terms of achieving their objectives. A great deal of what has been written and spoken about NW has been based on emotion and speculation rather than on scientific evidence and rational argument. The main reason for this is that an important part of the programme is the generation of positive thinking in order to encourage the public to believe that there is something that they can do to reduce crime and to motivate residents to take part in the schemes. There is an obvious danger that positive thinking will discourage positive criticism which in the long term might help rather than hinder the development of the programme.

It is an aim of this chapter to examine some of the facts about NW in order to provide a rational base from which to assess its future development and potential effectiveness. In particular the chapter will look at the number of schemes and their distribution around the country and variations in take up and support; it will describe the themes and variations in the design and implementation of NW programmes and consider some of the consequences of these variations. The chapter will conclude by reviewing some of the most urgent problems facing the design and management of NW programmes and will make recommendations on future developments.

NATIONAL TRENDS

There is some doubt about the precise number of NW schemes in the country. It was estimated last year in the official Home Office publication on NW *Good Neighbour* that there were currently over 60,000 NW schemes in the country and new schemes were being launched at a rate of over 1,000 a month (*Good Neighbour* 1988). More recent figures obtained by the *Observer* from the Home Office showed that there was a total of 74,000 schemes and that 7,500 new schemes had been created during the four-month period between March and June 1990 – a rate of over 600 new schemes a week.

The Home Office figures on the number of schemes are the only nation-wide source of information on the total number of NW schemes in the country. The article referenced above included the results of a small piece of research conducted by the *Observer* which suggested that the Home Office figures might not be reliable. Few of the forces contacted by the newspaper were able to confirm the Home Office figures relating to their force and some of them reported a total number of schemes less than half the Home Office estimates.

More reliable estimates on the spread of NW throughout the country are included in the report of the third British Crime Survey. The authors of the report estimate that at the beginning of 1988 14 per cent of households in England and Wales were members of a NW scheme covering approximately 2.5 million households (Mayhew *et al.* 1989). This compares with 25 per cent of the Canadian population (1988) and 19 per cent of the North American population (1984) (reported in Mayhew *et al.* 1989). The BCS report noted an expansion of awareness of NW during the periods between the second and third surveys with a change from just over half of the population having heard of NW in 1984 to 90 per cent of the population in 1988. It also noted that two-thirds of respondents not currently members of a NW scheme reported that they would be willing to join one if one were set up in their area.

273

REGIONAL VARIATIONS

Information on the regional distribution of NW schemes can be found in two BCS reports (Hough and Mayhew 1985; Mayhew *et al.* 1989) and in a smaller-scale study of nine police force areas commissioned by the Home Office Crime Prevention Unit (Husain 1988).

The most recent BCS report showed marked regional differences in the coverage of NW schemes throughout the country. The report divides areas into ACORN classifications of neighbourhoods (see Mayhew *et al.* 1989). Schemes were most common in 'affluent suburban areas' and 'high-status non-family areas'. Conversely, schemes were least common on 'less well-off council estates' and in 'poor quality older terraced housing' areas. Approximately 29 per cent of respondents in 'affluent suburban areas' reported that a scheme had been set up in their area compared with 10 per cent of respondents in 'less well-off council estates'.

The take-up rate of residents in areas with NW schemes tended to be a little higher (but not a direct correlation) in the high coverage areas than the low coverage areas. High coverage areas tended to have lower burglary risks and low coverage areas tended to have higher burglary risks (with some anomalies in this trend). This latter finding gives some support to the argument that NW is most frequently found in areas which need it least – although the statement is only true when 'need' is interpreted narrowly to mean burglary rate.

PROGRAMME DESIGN

One way of looking at programme design is to consider the structure and function of some of the most influential schemes. Another way is to examine details of a large number of schemes and to identify themes and variations.

Influential schemes

One of the first NW programmes in the United States was the Community Crime Prevention Program implemented in Seattle, Washington, in 1975. This scheme was particularly important because it was used as a model of NW and many schemes in both Britain and the United States developed programmes based on the broad principles of its design.

The Seattle programme comprised four main elements: (1) residential security inspection; (2) property marking; (3) block watches; and (4) information to the public to promote citizen awareness of their role in reducing residential burglary (Cirel et al. 1977).

The Block Watch component of the programme was administered by civilian organizers who arranged a launch meeting at the home of one of the block residents. The meeting included a discussion on crime in the areas, methods of contacting the police to report suspicious incidents, and security advice. At the end of the meeting the participants elected a Block Watch captain, who became the liaison person between the watch participants and the civilian organizers. The members of the scheme were encouraged to hold meetings at least once a year to maintain interest. The Block Watch usually covered between ten and fifteen households.

One of the first NW schemes in this country was the Home Watch programme implemented in the village of Mollington in Cheshire in July 1982 (Anderton 1985). The Mollington scheme comprised a comprehensive package of: (1) NW; (2) property marking; and (3) improvements of physical security in homes. The main aim of the scheme was to reduce house burglaries and other property crimes; the main method adopted to achieve this was to encourage neighbours to keep watch on one another's homes and to report suspicious incidents to the police.

Other early schemes included the Kingsdown Neighbourhood Watch Project launched in Bristol in February 1983 (Veater 1984). The Kingsdown Project was implemented by the police in response to demands from the public to do something about the problem of burglary in the area. The programme was based on a comprehensive package comprising: (1) NW; (2) a home security survey campaign; and (3) property marking. It also included two elements not found in the early North American schemes: a recruiting drive for special constables and an increase in regular foot patrols.

Perhaps the most important development in the early history of NW in Britain was the implementation of NW in London. An important aspect of the promotion of NW by the Metropolitan Police was the decision by the Commissioner of the force to launch the schemes on a force-wide basis. The first schemes were implemented in September 1983 and by the end of the year most divisional commanders had begun to implement NW programmes in their areas (Bennett 1987). The effect of the decision was not only to accelerate the rate of growth in London but also to signal to other forces that NW had the support of the largest police force in the country.

NW is described in the policy documents of the Metropolitan Police as 'a network of public spirited members of the community who become the eyes and ears of the police'. The principal aims of the programme were to reduce residential burglary and other street crimes such as robberies, auto thefts and criminal damage. The programme comprised a comprehensive package based on: (1) NW; (2) property marking; (3) home security surveys; and (4) community crime prevention and the development of environmental awareness.

Individual schemes were launched with a public meeting attended by the police and local residents. Residents attending the launch meeting were given a package containing a number of specially designed leaflets and booklets covering methods of crime prevention.

Once the scheme was established the crime prevention officer or local community constable would provide participants with NW and property-marking window stickers and instructional leaflets on how to mark property and report suspicious behaviour. In some areas a newsletter was produced by the police in collaboration with local residents. At a later date the police would arrange the erection of street signs if there was evidence of sufficient support among scheme members.

Design themes and variations

Another way of looking at the practice of NW is to identify some of the key elements which characterize typical schemes. An important structural element of NW which emerges from the literature is that it is often implemented as part of a comprehensive package. The typical package in both Britain and the United States has been: NW, property marking, and home security surveys – sometimes referred to in the United States as 'The Big Three' (Titus 1984). There is sometimes a fourth or fifth element of the package, which is an idiosyncratic programme devised to fit the needs of a particular police force or area. The Kingsdown NW scheme in Bristol, for example, included in its package a recruitment drive for special constables and increased regular foot patrols (Veater 1984).

An important difference which distinguishes British and North American schemes is that sometimes the package includes citizen patrols. The Neighborhood Fight Against Crime programme in Brooklyn, New York, for example, comprised a package which included: NW, operation identification, home security surveys, a telephone alert chain, and car and tenant patrols (DeJong and Goolkasian 1982). There is currently no police force in Britain which formally supports citizen patrols or which incorporates such patrols into a NW package, although there is evidence that there are some citizen patrols operating informally without active police support and independent of NW.

There is almost complete consensus in the published literature that the main aim of NW is crime prevention. The vast majority of programmes cite residential burglary as the sole or most important target crime of NW. Some programmes focus only on residential burglary while others list a number of offences which they hope NW will be effective in reducing. The list of other offences often includes: street robberies, auto thefts and criminal damage; or more generally: street crime and property crime. Apart from crime reduction, schemes often cite additional or supplementary aims, such as reduction of fear of crime, improvement in police–community relations, and improvement in crime reporting rates.

The size of NW schemes is also an important structural element. NW schemes in the United States are often based on small blocks of dwellings and include no more than twenty to thirty households. The size of the Block Watch units in the Seattle scheme, for example, varied from ten to fifteen households. Some of the largest NW schemes can be found in Britain: one of the early schemes in London was reported as covering over 3,000 households.

Typically, NW schemes are both public and police initiated. Schemes which are launched in the early stages of the life of the programme in a particular country tend to be police initiated, for example the Kingsdown scheme in Bristol and the early NW schemes in the Metropolitan Police District. Schemes which are launched at a time when the programme is well developed and active are more frequently public initiated. One reason for this is that public demand for the programme often quickly exceeds the ability of the police to launch them. Some police departments in the United States have continued to initiate their own schemes after the programme was fully developed. The police in Detroit, for example, continued to launch police-initiated schemes in order to promote NW in areas which were unlikely to generate public-initiated requests.

The number and type of public meetings held varies among NW schemes. Some schemes have public meetings which involve all of the residents participating in the programme while others have meetings which involve only the organizers of the scheme. In Britain, the most common type of meeting is the latter involving just the scheme organizers. In the United States, it is more common for meetings to be attended by all residents. The main reason for this difference is that a large proportion of North American schemes are Block Watches covering small numbers of households. Under this arrangement it is possible for meetings of all residents in the scheme to be held at the home of one of the block members rather than a hired hall.

VARIATIONS IN PARTICIPATION

Research into factors related to participation in community crime prevention programmes has indicated that there might be important social and economic differences among participants and non-participants. Shernock (1986) found in a study area in the United States that leaders of NW programmes were more likely than a control sample to be older, white, married and higher socio-economic status. Skogan and Maxfield (1981) found that participants were more likely than non-participants to be older, in high-income households, and living in a household with children. Whitaker (1986) found that participants in NW schemes were more likely than non-participants to be living in higher-income households, in larger households, home-owners, middle aged, and to have higher levels of educational attainment. Research conducted in London found that participants were more likely to be owners than renters and to live in houses rather than flats (Bennett 1989).

Research also shows notable attitudinal and behavioural differences between participants and non-participants. The findings of a five-year Law Enforcement Assistance Administration (LEAA) funded project on reactions to crime concluded that citizen involvement in anti-crime efforts was related to their involvement in neighbourhood and community groups (Skogan et al. 1982; DuBow and Podolefsky 1979; Lewis and Salem 1981).

Hope (1988) concluded from his analysis of data collected in the second British Crime Survey that support for NW was significantly related to being worried about becoming a victim of burglary and perceiving the probability of victimization as high. He also found that supporters of NW were more likely than non-supporters to perceive their neighbours as friends or acquaintances, to believe that

278

people in their neighbourhood helped each other, and to be involved with neighbours in mutual home-minding arrangements.

Bennett (1989) concluded that participants were more likely to be worried about being the victim of household and personal crimes and more likely to feel unsafe after dark. They were also more involved in their community and had more favourable perceptions of the police.

IMPLEMENTATION EFFECTIVENESS

The term 'implementation effectiveness' is used here to describe the extent to which the programme succeeds in achieving its implementation aims – that is the extent to which the structure and function of the scheme is established as intended.

One way of looking at the level of public activity within NW schemes is to look at what members do. The results of the third BCS show that more than three-quarters of all self-reported NW members said that they had put stickers or posters in their windows. A slightly higher proportion said their scheme had street signs declaring that their homes were in a NW area. These results are similar to those reported by Bennett (1987; forthcoming), who showed that almost all members of the two NW schemes investigated had placed NW stickers in their windows (although not all residents living within the area were members).

The authors of the third BCS report that only 4 per cent could be classified as 'inactive' in terms of use of street signs, production of newsletters or leaflets, and keeping members informed about how the scheme was working. Other research using other criteria of inactivity has produced less encouraging findings.

Husain (1988) found that property marking and house security surveys were taken up by only a minority of members in NW schemes. He reported that in only 41 per cent of schemes investigated had more than half of the scheme members marked their property and in only 13 per cent had more than half requested a home security survey.

The results of an evaluation of two NW schemes in London also found low levels of participation in terms of property marking and home securities (Bennett 1987; forthcoming). In addition, the research found disappointingly low levels of surveillance and rates of reporting to the police. It is a major aim of NW schemes that residents should become the 'eyes and ears' of the police. The report showed that fewer than half of respondents in one area and just over half in the other area said that they had deliberately looked out for anything suspicious over the last year. Only about half of respondents who had looked out for something suspicious actually saw something; fewer than half of these reported what they saw to the police.

OUTCOME EFFECTIVENESS

Most evaluations of the outcome effectiveness of NW in Britain and the United States have been conducted by the police involved in setting up the schemes. Almost all of these studies report that NW leads to a reduction in crime (mainly the offence of burglary). Unfortunately, a great deal of this research suffers from methodological weaknesses which undermine the value of the conclusions drawn (Titus 1984).

Stronger research designs have been included in evaluations conducted by independent researchers. The result of this research has been much less encouraging.

North American evaluations

A small number of evaluations in the United States have been conducted by independent researchers. One of the first evaluations was of the Seattle programme implemented in Washington in 1975 (Cirel *et al.* 1977). The evaluation was based on crime and public attitudes surveys conducted before the launch of the programmes and again after the programmes had been in operation for a period of time. The initial results of the evaluation were encouraging and showed a reduction in residential burglary for participants of over 60 per cent during the first twelve months of the scheme. However, a telephone survey of the same residents eighteen months after the implementation of the scheme showed that victimization rates had returned to their original levels and there was no longer any difference in victimization rates between participants and non-participants. The authors concluded that it is likely that the initial effects of NW programmes might wear off after a period of between twelve and eighteen months.

Another independent survey of a community programme which included programme elements similar to those of NW schemes was conducted in Hartford, Connecticut (Fowler *et al.* 1979). The researchers concluded that in the first two years of the scheme both burglary and robbery rates reduced significantly. A follow-up study two years later again showed that the effects were not maintained. The results revealed that two years after the launch of the programme there was no difference between the rates in the experimental area and in the city as a whole.

More recent evaluations in the United States have produced similarly discouraging findings. A study of NW in Washington DC found that in the first six months of the scheme the number of crimes increased, although in the second six months crimes decreased slightly (Henig 1984). However, during the same period crime in the city as a whole decreased by a greater amount than the programme area. No difference was found in changes in levels of crime between blocks that had active NW programmes and blocks that had inactive programmes.

The most surprising findings come from a recently completed evaluation of NW in Chicago (Rosenbaum 1987). The analysis was based on five community organizations which implemented NW schemes within their areas. The results of 'before' and 'after' surveys of experimental and control areas showed that in three of the five areas there was an overall increase in crime following the implementation of NW and in only one of the five areas was there a decrease in crime. Crime levels in Chicago as a whole remained stable over the same period.

In addition, the research revealed that in three of the five areas there was a significant increase in the fear of crime and in two areas the residents felt that crime had increased in their area. Optimism about the area declined in three areas and perceived likelihood of moving increased in two areas.

British evaluations

The only independent evaluation of NW in Britain which looks specifically at outcome effectiveness is the study of two NW schemes in London (Bennett 1987; forthcoming).

The London evaluation was based on a quasi-experimental design which involved conducting crime and public attitudes surveys before the implementation of the NW scheme and again one year later. Surveys were also conducted in two similar areas with no NW scheme to provide a control comparison. The effectiveness of the programme was judged by comparing the performance of the programme areas with the non-programme areas over the one-year experimental period.

The results of the evaluations showed no evidence that the programme resulted in a reduction in levels of victimization. In fact, in both NW areas the number of households victimized one or more times increased from the period before the launch of the scheme to the period after the launch. Reporting rates for household victimizations either remained stable or declined. the number of telephone calls for service made to the local police from residents in the NW areas declined in both areas. There was also no improvement in detection rates for the subdivisions hosting the NW schemes during the course of the experiment.

There were more encouraging findings in relation to changes in residents' attitudes and behaviour. Fear of household victimization reduced significantly over the experimental period in one of the programme areas but not in the other. In both areas there were significant improvements in perceived satisfaction with the neighbourhood. Involvement with neighbours in home protection improved significantly in one of the areas and sense of social cohesion improved significantly in the other.

The remaining findings were not so encouraging. There was no evidence that the NW schemes were associated with a reduction in fear of personal victimization or with a reduction in perceived probability of victimization or evaluation of police service. The most troubling finding was a significant reduction in both areas in contact with the police as measured by a question concerning the last time the residents saw a police-officer in the area.

CONCLUSION

There is no reason to believe that these negative findings are unique to specific evaluations which generated them. One of the important lessons that might be learned from this research is that the design and implementation of NW needs to be thought about carefully and honestly. In particular attention needs to be paid to the theory of NW and the crime preventing processes that the programmes are presumed to instigate.

The theoretical problems concern the general perspective on the causes of crime underlying NW programmes and in the specific assumptions about the way in which the component parts of NW are supposed to operate. It is important that some attempt is made to clarify more precisely the theory of NW. If it is decided that NW is dominantly a situational approach to crime then the opportunity reducing element of the programme should be emphasized. If it is decided that NW is dominantly a social control approach to crime then activities which strengthen informal social control should be promoted.

Attention should also be paid to some of the themes and variations in NW reported in this chapter. The themes of NW are important because they determine the shape of what is common and normal. The variations in NW are also important because they point to differences in design and implementation which might have potential consequences for residents living in these areas.

It must be discussed and decided whether the major themes of NW in this country are the right ones. One of the key themes that needs to be considered is the accepted design of the programme. It is possible that the current level of activity of most schemes is inadequate to invoke the necessary social processes to reduce crime. Perhaps more thought might be given to whether more could be done to reduce opportunities for crime apart from simply encouraging residents to become the 'eyes and the ears' of the police. Similarly more thought might be given to whether more could be done to increase informal social control apart from holding periodic meetings of co-ordinators or (less frequently) programme participants.

Some thought might also be given to the variations in NW schemes. The differential take-up of NW is an important issue. Research shows that NW is most frequently enjoyed by residents in affluent areas containing little crime. It is not necessarily the case that differential take-up is a bad thing as there might be good reasons why NW is more suitable for some areas than others. Nevertheless, geographical variations in the distribution of NW need to be carefully considered and the full implications of it needs to be assessed.

Some attention might also be paid to variations in the design of NW programmes. It has been shown that schemes vary in size from just a few dwellings to whole housing estates. Some schemes are fairly active in terms of participation while others are fairly inactive. It is unknown what effect these variations have on the potential success of the programme in achieving their objectives or on programme maintenance over time.

The future of NW lies in the critical appraisal of its past. The last quarter of a century of criminological research has shown that it is not easy to reduce crime. An important lesson to learn from this is that simple solutions are unlikely to produce major gains. In order to maximize the potential impact of NW it is necessary to develop and refine the programme design in order to determine from experience rather than guess-work what is and what is not effective. This would require a process of continuous evaluation and a management system which acknowledges and responds to both programme successes and failures.

NOTE

1 The phrase 'neighbourhood watch' is used in this chapter as a general term to cover a wide variety of schemes which are variously referred to as 'Home Watch', 'Community Watch', 'Farm Watch', and so on.

REFERENCES

Anderton, K.J. (1985) *The Effectiveness of Home Watch Schemes in Cheshire*, Chester: Cheshire Constabulary.

Bennett, T.H. (1987) *An Evaluation of Two Neighbourhood Watch Schemes in London*, Report to the Home Office, Cambridge: Institute of Criminology.

—— (1989) 'Factors related to participation in neighbourhood watch schemes', *British Journal of Criminology* 29 (3): 207–18.

—— (forthcoming) *Evaluating Neighbourhood Watch*, Aldershot: Gower.

Cirel, P., Evans, P., McGillis, D. and Whitcomb, D. (1977) *Community Crime Prevention, Seattle, Washington: An Exemplary Project*, US Department of Justice, Washington, DC: Government Printing Office.

DeJong, W. and Goolkasian, G.A. (1982) *Neighbourhood Fight Against Crime: The Midwood Kings Highway Development Corporation*, National Institute of Law Enforcement and Criminal Justice, US Department of Justice, Washington, DC: Government Printing Office.

DuBow, F. and Podolefsky, A. (1979) *Citizen Participation in Collective Responses to Crime*, Evanston, Ill: Center for Urban Affairs.

Fowler, F.J., McCalla, M.E. and Mangione, T.W. (1979) *Reducing Residential Crime and Fear: The Hartford Neighborhood Crime Prevention Program*, Washington, DC: Government Printing Office.

Good Neighbour (1988) 'Comment', *Good Neighbour* 6: 3.

Henig, J. (1984) *Citizens Against Crime: an Assessment of the Neighborhood Watch Program in Washington, D.C.*, Washington, DC: George Washington University.

Hope, T. (1988) 'Support for Neighbourhood Watch: a British Crime Survey analysis', in T. Hope and M. Shaw (eds) *Communities and Crime Reduction*, London: HMSO.

Hough, M. and Mayhew, P. (1985) *Taking Account of Crime: Key Findings from the 1984 British Crime Survey*, Home Office Research Study 85, London: HMSO.

Husain, S. (1988) *Neighbourhood Watch in England and Wales: A Locational Analysis*, Crime Prevention Unit Paper 12, London: Home Office.

Lewis, D.A. and Salem, G. (1981) 'Community crime prevention: an analysis of a developing strategy', *Crime and Delinquency* 405–21.

Mayhew, P., Elliott, D. and Dowds, L. (1989) *The 1988 British Crime Survey*, Home Office Research Study 111, London: HMSO.

Observer (1989) 'Home Office inflated watchdog figures', *Observer*, 27 August, London: Observer Newspapers.

Rosenbaum, D.P. (1987) 'The theory and research behind neighborhood watch: is it a sound fear and crime reduction strategy?', *Crime and Delinquency* 33 (1): 103–34.

Shernock, S.K. (1986) 'A profile of the citizen crime prevention activist', *Journal of Criminal Justice*, 14: 211–28.

Skogan, W.G. and Maxfield, M.G. (1981) *Coping with Crime: Individual and Neighborhood Reactions*, Beverly Hills, Calif.: Sage.

Skogan, W.G., Lewis, D.A., DuBow, F. and Gordon, M.T. (1982) *The Reactions to Crime Project: Executive Summary*, Washington DC: US Department of Justice.

Titus, R. (1984) 'Residential burglary and the community response', in R.V.G. Clarke and T. Hope (eds) *Coping with Burglary*, Boston, Mass.: Kluwer-Nijhoff.

Veater, P. (1984) *Evaluation of Kingsdown Neighbourhood Watch Project Bristol*, Bristol: Avon and Somerset Constabulary.

Whitaker, C.J. (1986) *Crime Prevention Measures*, Bureau of Justice Statistics, Special Report, US Department of Justice, Washington, DC: Government Printing Office.

15

INITIATIVES IN POLICING LONDON'S BRIXTON SINCE THE 1981 RIOTS

David Mitchell

Lord Scarman in his report on the Brixton disorders of 10–12 April 1981 called for the establishment of 'statutory liaison committees, or other appropriate consultative machinery' (Scarman 1981: para. 5.71; see also Morgan and Maggs: 1984). The statutory framework was put in place through the Police and Criminal Evidence Act (PACE) which progressed through Parliament during the summer of 1984. PACE stated that

> Arrangements shall be made in each area for obtaining the views of people in that area about matters concerning the policing of the area and for obtaining their co-operation with the police in preventing crime in the area.
> (PACE 1984: Section 106(1); see also Morgan 1989: 220–1)

The Act, however, did not require that the arrangements made were of any specific form. Instead a series of Home Office Circulars outlined the suggested form of consultation, namely that of formal consultative groups (Fyfe 1989a; Fyfe 1989b; Morgan 1987b; Morgan 1989). The series of circulars also emphasized that the nature of the consultative process would be different from place to place, although with the general aim of 'sustained improvement in relations between the community and the police; and to promote agreed solutions to local problems' (Morgan and Maggs 1985a: 17).

Early work by Morgan and Maggs (1984; 1985a) focused on the implementation of Police Consultative Committees (PCCs) across the provincial police authorities of England and Wales, specifically excluding London (see Chapter 16 by Fyfe in this volume and Fyfe 1989a for research on consultation in London). Despite the pressure exerted by Home Office Circulars there has still been the flexibility for local consultancy groups to emerge in response to specific local needs. Within London the consultative process has been based around boroughs, leading to very specific problems which will be examined below (Fyfe 1989a; Morgan 1989). In this chapter I shall focus on one local police community consultative group, the Community/Police Consultative Group for Lambeth, and examine how the practice of consultation has operated. Specific commentary will be given on how this group has effected the dialogue between police and community within the Brixton area of Lambeth.

CAVEAT EMPTOR

Given the highly charged political nature of the policing debate in London it is perhaps worth giving a few warnings about the choice of Lambeth, and Brixton in particular, for analysis.

Three warnings should be issued to the reader. The first concerns the spatial organization of the Metropolitan Police. Prior to the restructuring of the Metropolitan Police force in 1985 the force was organized as a series of twenty-four police districts, which were either coterminous (in ten cases) or aligned with borough boundaries. For example, 'L' district of the Metropolitan police was coterminous with the borough of Lambeth. Therefore, policing policies were not unique to Brixton but were within the framework of the policing of Lambeth as a whole. Subsequent to the reorganization of the Metropolitan Police in 1985, London was, and still is, subdivided into eight policing areas. Lambeth and Brixton come under the organizational umbrella of '4' area, which stretches from Kennington in the north to Croydon and Banstead in the south (GLC 1986). Currently, issues of the policing of Lambeth cannot be totally divorced from those of '4' area.

The second warning relates to the volume of change and the number of policing initiatives that have taken place in Lambeth since the 1981 riots. Initiatives have been taken by both the police and the local council to improve the policing of the area, and the dialogue between police and community. The focus in this chapter is on three initiatives: the community/police consultative group, Lay Visitors and community involvement officers.

The final warning issued to the reader is on the subject of the perpetuation of negative stereotypes. Brixton has been the scapegoat in a great deal of analyses of the problems associated with policing London, and the Black communities of London in particular.

There is thus a danger, that when highlighting Brixton for study, of helping to perpetuate politically negative stereotypes of Brixton when it would be more constructive to try to work against those stereotypes (see Keith 1987).

VOICES FROM THE COMMUNITY?

The account I present here is based around a number of data sources. These fall into four groups: personal observation, informal interviews, formal interviews and documentary research. Each of these is discussed in turn below.

Personal observation

From the beginning of 1987 I carried out field research as part of a wider research project into the causes and consequences of rioting in Britain's inner cities in the 1980s. The style of the field research was essentially qualitative (see Burgess 1982; Eyles and Smith 1988; Hammersley and Atkinson 1981), and stressed the active participation of the researcher. This field research took me to Brixton on many occasions, where I made many friends in the community and in the police force. During 1987, 1988 and the first half of 1989 I worked in Brixton, meeting people from all sections of the community. In 1988 I served on the committee of one of the local youth clubs in Central Brixton, helping wherever possible to contribute to the running of the club and to the well-being of its members. Over the time period I worked in Brixton I established a wide variety of contacts with a broad spectrum of people in the community. These people ranged from youths from the Moorlands estate in Central Brixton, local residents' associations, church leaders, and junior and senior police-officers. Some of the comments and conclusions contained herein come from my observations from talking to, and working with, these individuals as opposed to directly questioning them on specific topics.

Informal interviews

I conducted a series of informal semi-structured interviews about the consultative process in Lambeth, and the way that it was viewed by the community. Such interviews were carried out at irregular intervals throughout the 1987–9 period. The interviewees were a range of community and youth leaders from around Lambeth, whom I met while working with one youth club in Brixton. Similar interviews were also conducted with community involvement, youth and community section and crime prevention officers from the police.

Formal interviews

A series of more formal interviews were conducted with senior police-officers, usually of the rank of Inspector or above who had responsibilities for aspects of the consultative process. The interviews were more structured than those mentioned above, with drafts of questions presented to the officers several days in advance.

Documentary research

Documentary research was also undertaken on the formal constitution of the consultative group and Lay Visitors scheme described on pp. 294–8, based on the minutes of meetings.

THE LAMBETH COMMUNITY/POLICE CONSULTATIVE GROUP

The Community/Police Consultative Group for Lambeth (CPCGL) was the first consultative group to be established in Britain, on 3 March 1982 (Cansdale 1983). The importance of the establishment of a consultative group in Lambeth was demonstrated by the participation of the then Home Secretary William Whitelaw in the initial discussions.

This section of the chapter looks at the goals, membership and operation of the CPCGL and comments upon their relevance for the local community. Three objectives are set out in the CPCGL constitution. These are that the group aims to foster better relations between the police and the community (special reference is made to Black and ethnic minority communities), to develop better crime prevention and to maintain 'a peaceful community in the Borough of Lambeth' (Community/Police Consultative Group for Lambeth 1984). In terms of the guidelines circulated by the Home Office there is broad accord between the specific CPCGL aims and the purposes of consultation generally.

Brogden *et al.* (1988) describe the membership of police consultative committees as generally being representatives of the police, the police authority, constituent councils, and voluntary, statutory and community groups. The membership of the CPCGL is 'open to all bona, formally constituted local groups, representing a significant number of people in the Borough' (CPCGL constitution). In practice this constitutional principle translates to what is often known as an open forum membership (Morgan and Maggs 1984).

In the CPCGL various categories of membership exist, including Lambeth Members of Parliament, Lambeth councillors (appointed by the various political groups), police representatives and community and voluntary organizations. It is a specific policy of the group that community and voluntary groups should be greater in number than any other category, and that this balance should be reflected in the voting rights by allocating one vote to each member. There are a huge variety of affiliated community and voluntary groups. Table 15.1 shows a broad classification of these groups.

Table 15.1 The composition of the CPCGL

Afro-Caribbean groups	5
Asian groups	3
Business groups	2
Co-opted members	6
Local government groups	4
Members of Parliament	2
Miscellaneous	5
Religious groups	10
Sector working parties	7
Tenants' associations	14
Youth clubs	2

Source: Community/Police Consultative Group for Lambeth 1987

The categories defined are necessarily crude, especially in view of the fact that each organization represented may perform functions which would place it in more than one of the categories. The sector working parties referred to are smaller-scale initiatives and tied to each of the four police divisions within Lambeth (Brixton, Clapham, Kennington and Streatham) and designed to provide detailed analysis of policing requirements in their divisions.

It is evident from Table 15.1 that the two largest groups are religious groups and tenants' associations. Two points need to be made here. First, that it would be difficult for any collection of groups to 'monopolize' the CPCGL, because of the wide range of groups represented. This ensures that the CPCGL is broadly representative. Second, the low number of groups within the Afro-Carribbean category is not necessarily a reflection of a low Afro-Caribbean participation, since their interests are also represented within other categories, for example the Brixton Domino Club.

The question of membership raises the issue of representation, that is does the CPCGL represent the broad interests of all those within the community or is it sectional? Other authors have noted that police/community consultative group 'members are disproportionately male, middle-class, middle-aged, "involved" members of the community. They constitute a sort of "great and good" locally' (Morgan 1989: 229; see also Maguire and Vagg 1984; Fyfe, Chapter 16 in this volume). The picture of representation presented in Lambeth is very much in accord with this general image. The focus of attendance on the involved members of the community has been high. A quote from a local Black youth, in his early twenties, from the Moorlands Estate demonstrates one perception of how such participation is viewed. He stated that: 'They've [the CPCGL members] got their fingers in every piece of pie; in the TA [Tenants' Association], youth club committees, everything. They just enjoy the feelin' of being important, of being in the middle of things'.

As Morgan (1989) also notes, there is a tendency for members of police/community consultative groups to have had little adversarial contact with the police. Within the CPCGL it was recognized that the participation of Black youths within the CPCGL was essential. A quote from one Black youth leader from Central Brixton illustrates some of the general feelings held by Black youths about the CPCGL.

> The group [the CPCGL] does a lot of things that the youth couldn't do on their own. We get to hear what the top dogs say and to get to have a go at them, when things go wrong.

This quotation highlights two things. First, that one of the features of the CPCGL that is appreciated by Black youth is access to high-ranking officers. Second, that there are occasions when there is direct participation in the CPCGL by Black youths, although on only a few occasions have Black youths made vocal contributions to the CPCGL.

One of the dangers that Cochrane (1986) points to is that those who previously represented sections of the community can become marginalized if they appear to give too much support to the police. Given the long-running conflicts between the Black community and the police within Lambeth, and specifically Brixton, there was a great danger of Cochrane's scenario taking place (London Borough of Lambeth 1981; Scarman 1981). Morgan also indicates cases where 'the "community leaders" on the consultative committees were subsequently disowned by other neighbourhood "spokespersons"' (Morgan 1987a: 37). A certain amount of the marginalization that the above authors describe has occurred within Lambeth. In one case a senior police-officer said that 'Leader X has completely lost the respect of the kids on Moorlands through speaking up at the consultative group'.

As well as looking at the groups of people represented in the CPCGL there is also a need to look at the specific geography of representation. At the micro-scale there has been a noted conflict over police consultative committees. As Morgan (1987a) notes: 'they do not represent a community [and] sub-divisions seldom coincide with self-identifying physical neighbourhoods or communities' (1987a: 42). The CPCGL actually copes well with this potential scale conflict, with its territorial basis being Lambeth, an area with well-perceived boundaries. Through the sector working parties the CPCGL is able to operate at a smaller spatial scale. At CPCGL meetings a chief superintendent from each of the four divisions in Lambeth reports to the group.

The mode of operation of the CPCGL has had an effect on the degree to which it is representative of the community at large. Meetings of the group tend to be dominated by the reports of the four chief superintendents (one from each Lambeth division) usually outlining crime statistics and clear-up rates for their division. Here the workings of the group seem to fall into line with those for similar groups around the country (Morgan 1989). The work of the CPCGL is, in practice, subdivided into subcommittees which examine police behaviour and complaints, conditions in police stations (the Lay Visitors, see pp. 294–8), crime prevention, racial attacks, policing training and recruitment and legislation. The process of subdivision and the reporting back to the main CPCGL that ensues is one feature which seems to perpetuate the lack of integration of many Black youths into the formal consultative process. Meetings tend to be structured into a series of reports, from the police and from the various working parties, leading to a feeling among Black youths that much of the decision-making and power of the group is inaccessible to them. Many Black youths (and others) feel that, because of the structure of the meetings, they are unable to contribute and would not bring problems over policing to the group. Part of the lack of participation is due to a perception that the group has no real powers of sanction over police behaviour. To quote one Black youth from Kennington: 'We wouldn't go near the consultative group, it's just not for us. Even if we did complain to them about something that the police did then what could they do?'

Having given an analysis of the individual topic of goals, membership and operation an overall assessment should be made of the CPCGL. This assessment will be within the framework that Morgan (1989) laid out. Morgan outlined three hypotheses on the function on police consultative committees: the marginalization, legitimation and penetration hypotheses. The marginalization hypothesis is where only those who are advantaged by the consultative committees or have sympathies with the police participate in PCCs. Those who do not participate within the PCCs, and who are critical of the police, can then tend to be progressively closed out of the consultative process. The legitimation hypothesis occurs when the members of the community who attend PCCs begin to learn about the operation of the police and adopt the ideals of the police force. In other words everyday police administrative constraints can be given too much importance and can govern what operational policies the PCCs wish to see implemented. The penetration hypothesis happens when there is a progressive blurring of the distinction between formal and informal policing, with the PCCs beginning to be associated with the day-to-day policing of the local area.

In terms of Morgan's own work he found substantial support for the legitimation and marginalization hypotheses, but none for the penetration hypothesis. The overall impact of the PCCs was to increase contact between those 'involved' members of the community and the police, and permit those members of the community to learn more about the activities of the police. With specific respect to Lambeth it appears that processes of marginalization and legitimation have both been in operation. The group who appear most marginalized are Black youths, who, despite attempts to integrate them, still feel that they have little role in the forum of the consultative group. Some form of legitimation has occurred through the structure of the consultative group meetings being focused around police reports, enabling the members of the group to learn about the operational dynamics of policing in Lambeth. Part of the lack of confidence that many Black youths express in the CPCGL has been through the fact that they see the meetings as being oriented around police, rather than community needs. In this way they see the CPCGL as being an extension of the activities of the police.

THE LAY VISITORS SCHEME

The Lay Visitors scheme is the second initiative affecting the dialogue between community and police to be considered. The initial idea of Lay Visitors to police stations was brought to public attention when Lord Scarman commented on the possibility of introducing such a scheme in his report (Scarman 1981: para. 7.7–7.10). The Lay Visitors scheme in Lambeth was one of the first pilot schemes across the country, the others being in Greater Manchester, West Midlands, South Yorkshire, Humberside, Leicestershire and Cheshire. The Lambeth scheme was initially negotiated between the Home Office, the CPCGL and the Metropolitan Police during 1982 and 1983, finally being introduced in January 1984. As Brogden *et al.* (1988) state the schemes enable 'members of the community to observe, comment and report on the conditions under which persons are detained at police stations and the operation in practice of statutory and other rules governing welfare' (1988: 176).

Lay Visitors are appointed for a period of three years by the Home Secretary and operate under a set of guidelines issued by the Home Office. Visitors are given a basic training course, at the Metropolitan Police Training College at Hendon, so that they are aware of the relevant sections of the law. In Lambeth, in addition to the centrally provided guidelines, there are locally negotiated arrangements as to the operation of visits which were introduced after a dispute between the Lay Visitors and the police in Lambeth which developed in December 1987. The dispute, basically, concerned the way in which the visits were conducted, and in particular on the role of the police-officer accompanying the visitor.

The Panel of Lay Visitors meets once a month to discuss the visits that they have made and any matters arising from them. At least twice a year the panel must report to the Home Secretary, with a similar report given at least four times a year to the CPCGL.

The Lay Visitors have the freedom to chose the timing of visits and their location, although they are required to visit each of the five police stations in Lambeth at least monthly, with a rule of thumb limit of around fifty to sixty visits per year suggested as appropriate by most parties. However, the timing of the visits and the relative choice of police stations to visit is expected to be chosen so that no substantial burden is placed upon the operational staff of any one police station. In addition there is the potential for visits to occur when such a visit would calm a particular perceived tension within the community. For example, provision was made available through the community liaison office and the deputy assistant commissioner for visits to be carried out during and subsequent to the anti-poll tax demonstration which took place in Lambeth on Thursday 29 March 1990.

Following a disagreement over the conduct of the visits in the Brixton area (mentioned above), which happened in December 1987, a set of local guidelines were constructed. A police-officer is required to be present with the Visitors, for reasons of safety. The Visitors are permitted to speak to the detained person, subject to the agreement of the detained person and to the condition that conversations should take place in sight of the escorting officers, but out of that officer's hearing. Exceptions to this rule are when the detained person is being held incommunicado or when an officer of the rank of superintendent or above has decided that Lay Visitors should be excluded. If an interview is already in progress then that interview cannot be terminated for a visit to begin. A general agreement also exists that if the Visitor knows the detained person then the Visitor should declare that he or she knows the detained person and withdraw from the visit. The Visitor may be able to return to the visit, subject to the approval of other interested parties.

In terms of the subject matter of the visits the Visitors are required to stick tightly to their remit, broadly the condition of the detained person and his or her treatment. Any other areas discussed, especially if they refer to the charge that the detainee is facing or may face, may present legal problems of *sub judice*. After the interview the Lay Visitor is required to fill in a form reporting on the visit, one copy of which is given to the chief superintendent and another copy sent to the secretary of the Panel of Lay Visitors. Where there is concern over specific matters there are specific provisions for the immediate follow-up of the problem. For example, if a prisoner's clothes had been removed for forensic examination and the prisoner had not been provided with adequate substitute clothing then the Lay Visitors could request that such clothing be made available immediately.

When commenting upon the success or failure of the Lay Visitors scheme there are three topics which need to be discussed. These are whether the Lay Visitors scheme fulfils its stated aims, if the scheme has wide support in the community and if the scheme has improved the dialogue between police and community. Each of these three topics will be considered in turn.

In terms of the potential for improving dialogue between police and community, the stated aims of the Lay Visitors scheme are clear. These are to examine the conditions under which prisoners are detained in police stations in Lambeth. The Visitors have been successful in achieving this. Burney (1985), a Lay Visitor in Lambeth, notes that each of the five police stations within Lambeth have around one visit per week. As well as visiting *per se*, the Visitors have a responsibility to make constructive criticism, and have done so. The provision of extra clothing for prisoners whose clothes have been removed for forensic purposes has been one area where criticism has been made to good effect. However, the consequences of the Lay Visitors scheme are potentially much wider, in the breaking down of myths that surround the events taking place in police stations.

Burney (1985), commenting on the Lay Visitors scheme in Lambeth, stated that

> perhaps the most hopeful sign is the initial success of the drive to recruit young black people onto the panel. Since so many police prisoners in Lambeth are also young and black, their presence has wide implications, not all of them foreseeable.
>
> (Burney 1985: 240)

The picture in the Merseyside pilot project was rather different, with Walklate (1986a; 1986b) commenting that the largest group of participators in the project were white, middle-aged, professional people who had little adversarial experience of the police. From the differences between the two pilot projects it would seem that the Lambeth Lay Visitors scheme has attracted some degree of local popular support. Burney (1985) also notes that the composition of the Lambeth Lay Visitors contained a good mix of different occupational, ethnic and gender groupings. The information presented to me during formal and informal interviews with people in the

Brixton community was slightly different, with a less rosy picture being presented by most. A portrayal of the Lay Visitors as being a gathering of 'the good and the great' was common.

As Morgan and Maggs (1985b) note there is a great danger of events inside police stations generating horror stories about police misconduct, simply through the activities taking place away from direct public scrutiny. Burney (1985) similarly notes the potential for exaggerated accounts and points to the potential of the Lay Visitors scheme for avoiding the development of such horror stories (see also Brogden *et al.* 1988). From the interviews that I have carried out it would seem that there is now much greater confidence in the safety of prisoners in police stations than there was before the scheme began.

Although many in the community know little of the detailed workings of the Lay Visitors Panel there appears to be a high degree of trust in them. It is assumed by many Black youths that simply because the police appear not to appreciate the intrusion that the Lay Visitors can sometimes cause that the Lay Visitors must be performing a valuable task. As one youth explained: 'If they [the police] complain about them [the Lay Visitors] being a pain in the neck then they [the Lay Visitors] must be getting something right'.

A common fear held among many youths is that their friends held in custody will be beaten up or abused by the police. For such people the Lay Visitors scheme is useful, since the availability of access to detained people within the police station proves that the detained person is being held in reasonable conditions and has not been assaulted by the police. Thus a common set of fears can be put aside and genuine dialogue between this section of the community and the police can be brought that much closer.

For other sections of the community the Lay Visitors scheme has two separate interpretations. The first is that the Lay Visitors scheme represents little more than another layer of police and/or local state bureaucracy. Here the criticism has been based upon the formalized procedure that has developed. The disquiet felt by some over the dispute of December 1987 is an example of this first perception in operation. 'If they were really in it for the welfare of the people held in the station then they wouldn't have bothered with walking out; they would have just got on with it' is an opinion that has been expressed by both police officers and community leaders. The second interpretation views the Lay Visitors scheme as being an extension to the CPCGL, providing a degree of consultation but operating in a different forum – on police territory. The idea is that the Lay Visitors scheme provides a form of consultation, since it provides part of the framework for account-ability (Scarman 1981: para. 5.55–5.58). It is a form of consultation that allows the community to see, through impartial representatives, that the powers of the police, within the police station, are not being abused. Such a function is more important than might appear at first since it constitutes consultation on police territory, emphasizing that consultation is a two-way process not merely something that the police come to meetings to carry out.

COMMUNITY INVOLVEMENT OFFICERS

One of the most visible moves made by the Lambeth police has been the introduction of the community involvement officers. A community involvement office exists in each of the four divisions within '4' area, with responsibilities for increasing police involve-ment within the community. The responsibilities for increased involvement can be seen through such diverse events as arranging coach trips to the Royal Tournament, football tournaments, attend-ing community liaison meetings and helping women's groups. In other words the responsibilities are very varied but can generally be classified as encouraging better relations between the police and all sectors of the community, simply through increasing contact between the police and the community.

Having been based in a central office (in Brixton) until 1987 the community involvement officers were devolved to the four divisions. The move placed more personal stress on each officer since each had more individual responsibility, as opposed to having part of a collective responsibility. As a result of this devolvement, and the concomitant stress, there were a series of moves from the individual offices to increase their staff. This resulted in an increase in the overall numbers of officers in each office to approximately double its original number of one or two officers per division. For example, the Brixton Community Involvement Office increased to its current size of three police constables and one civilian admini-strative assistant.

Within the Lambeth police force there is a very varied attitude towards the community involvement officers (CIOs) and their work. Many have pejorative interpretations of the CIOs, since they are associated with soft policing, which clashes with the more traditional 'cuff 'em and stuff 'em' views of policing that are still held by some, both in junior and senior ranks. However, the respect given to the work of the CIOs has gradually begun to increase. One example of this growing respect is through the work of PC Bob Opray, one of the community involvement officers for the Brixton division. In 1988 PC Opray was awarded the BEM for his services to community policing, with the initial suggestion for the award originating from his colleagues and supervisors.

In terms of those within the Brixton community that I have spoken to there is a great deal of respect for the work of the CIOs, with enormous respect for particular officers. Such respect (from my observations) was not based around sectional interests. Black youths in the area shared the respect, one claiming that 'Bob's [PC Opray] a good guy, he treats us like human beings, not like some piece of shit'. Much of the respect given to the CIOs is based on the perception that they are helping to improve the facilities for the community, with motivation not being through self-interest. The help, both financial (through the discretionary budget of the CIOs, allocated from the '4' area community liaison office funds) and personal that the CIOs provide with a whole range of community projects has brought them closer to most parts of the community than other police-officers. The relationship, from my experience in the area, seems to be with all ethnic groups in the area, and with all age groups. One example of this trust in practice that I witnessed was a Black youth, about 18 years old, come into the Brixton police station, asking to see one of the CIOs. He said that he had been involved in a burglary a few weeks previous and wanted to give himself up, but to one of the Brixton CIOs and not to 'one of the local jokers'. It is also quite common for CIOs to be co-opted on to the committees of various social clubs within the district, where they often play a very active role.

However, care should be taken when assessing the overall impact of the CIOs. Much of the respect they gain is based upon personal admiration of their personal efforts, rather than for their status within the force as CIOs. It seems that the personal commitment to the topics that the CIOs work on which are important in gaining the admiration of the community. For example, a former female CIO within the area worked hard to highlight the problems of domestic violence. Other CIOs have likewise gained personal admiration through the perceived local importance of specific projects that they have worked on.

Community involvement officers come within the category of community policing, which also includes patrol, preventive work to tackle recurrent problems and information gathering relevant to crime detection (Weatheritt 1987). She sees the work of the CIOs as essentially worthwhile, though doubts whether the role that they now fulfil was envisaged when the units were first set up. Within Lambeth the situation is somewhat similar to this general pattern, although some differences exist. During informal interviews an image of CIOs being an adjunct to formal policing emerged. In addition those interviewed saw the CIOs as being a valuable contribution to improving community–police relations. Such views have been expressed by broad sections of the community, representing those from all interest groups. The idea of the CIOs being outside mainstream policing has obviously encouraged the establishment of such views and has probably improved their image within the community. The CIOs are also perceived by many to be participating within the community without wanting anything in return.

In terms of the criticism of Weatheritt (1987), namely that CIOs often have little direction in their work the Lambeth experience is in accord. The CIOs in each of the four divisions, although given general direction by the central community liaison office (CLO), are largely free to pursue the strategies that they think will help improve relations between police and community. However, the CIOs are also used by the office in the roles of information gathering and community intervention, especially at times when there is high tension in the area. For example, when a demonstration or protest march is planned in the area the CIOs are used to 'test the temperature' of the area before the outset of the demonstration, by talking to people from different sections of the local population. During such times the CIOs operate within what is termed the community intervention office, one of the functions of which is to allay any community disquiet about the policing of the demonstration. However, it should be pointed out that such situatons are rare and that the everyday operation of the CIOs is more akin to the pre-emptive roles described above.

CONSULTATION: WHO FOR?

Consultation is an active process and involves an evolving dialogue between police and community. Each of the three different initiatives that I have outlined above has a different effect on the overall consultative process. Each generates different perceptions within the community, has a different focus, a different audience and operates at different social and spatial scales.

As Morgan (1989) and Morgan and Maggs (1984) noted police community groups such as the CPCGL have raised levels of formal consultation between the police and the community, with backing from many strands of community groups. However, Morgan (1987a; 1989) has raised the issue as to the extent to which the consultative groups are representative of all aspects of the community and representative of those most in conflict with the police in particular. Within Lambeth the above descriptions of the workings of the consultative group have aimed to show that there are substantial sections of Black youth who feel that they have nothing to gain from the consultative group, and who would not participate therein. Indeed it seems as though the formal structure of the CPCGL, with its division into subcommittees and working parties, increases the sense of alienation these youths feel. Such a subdivision of the group leads to meetings becoming reportage in nature and takes away the possibility of the group being a forum where genuine dialogue can take place.

The Lay Visitors schemes across the country have run into similar problems with its members being accused of being the great and the good (Brogden et al. 1988; Morgan and Maggs 1984; Morgan 1987a). But, as Burney (1985) notes, the Lay Visitors scheme in Lambeth has largely been successful in avoiding this issue of poor representation. However, as indicated above, even were there poor representation of Black youths on the panel of Lay Visitors, there would still be a role for them. This role takes the form of being able to demonstrate the welfare of prisoners detained in police cells.

There are variations in the territorial extent to each of the three initiatives. The consultative group is viewed by many as a token form of consultation that the police come out to perform, only to return to the police station afterwards with the same opinions. The Lay Visitors scheme provides a form of consultation with a different territorial extent, whereby representatives of the community consult with the police on police territory. It is the fact that this form of consultation is intrusive upon police practice that gives it its value for many Black youths. In contrast to the previous two initiatives is that of the CIOs. Here consultation is carried in a number of different locations; on the street, in the social club, on the estate.

Consultation at this smaller scale can be seen as a process of more genuine dialogue, between those who are policed and those who police them.

All three initiatives thus have very different spatial resolutions. They also have very different social implications, in that they each have a unique audience. Morgan (1987a), with reference to police/community consultative groups, thought that it was doubtful if 'more than a few percent of the public have heard of their existence' (1987a: 231). Within Lambeth those who participate within the group, or who have contacts with it, are convinced of its value, noting that

> It [the CPCGL] has been a part of the evolution of events which has led to greater confidence in the Borough [of Lambeth], more co-operation with the police on the part of ordinary citizens ... and more imaginative and community-conscious policing.
>
> (CPCGL 1984: 5)

However, few people I have spoken to in Lambeth have been aware of the operation of the CPCGL, have been able to tell me about its membership or meetings, or about the initiatives that it has carried out. This does not imply that the CPCGL has not campaigned successfully on individual issues, merely that it is not a broad-based community group. Of those who were able to inform me about the CPCGL their main area of knowledge was on the 'national role' of the CPCGL, not on its specific Lambeth role. Here the contribution of the CPCGL towards changing the police training programme on 'Human Awareness' is quoted as an example.

The consultative group operates through communication between senior officers within the police hierarchy and formal community representatives. However, as has been pointed out above, there is some doubt as to whether those community representatives carry the opinions of all of Lambeth. Indeed, it has been argued above that participation within the consultative group has resulted in some community representatives being labelled as no longer representing the interests of the community.

There have also been some social implications to the establishment of the Lay Visitors scheme in Lambeth, though these cannot be fully divorced from the territorial effects discussed above. Access to police territory allows communication between the police and the community to take place on a different footing, with the police made to feel uncomfortable by the presence of members of the community. The fact that the Lay Visitors have freedom to choose where and when to make their visits also leads to a perception that the community are in charge of the process, rather than the police.

The community involvement officers have different social implications. They provide access to low-ranking officers, usually police constables, to all members of the community, not just those who claim to be community representatives. In other words the CIOs provide access to those who do the policing for those who are policed. There is no aura associated with police rank to contend with and dialogue can proceed on diverse, and relevant, topics.

This chapter has, then, tried to argue two things. First, that the three consultation initiatives outlined operate at different spatial scales and with differing social impact. Some initiatives are irrelevant to local neighbourhoods or to those in conflict with the police (Morgan 1989) while others such as the community involvement officers have more relevance for those being policed. Second, it has been argued that consultation is about people, not about committees. Consultation concerns the establishment of a genuine dialogue between the police and the community. It is important to realize that consultation is inherent in all police work, not just through the formalisms of consultative meetings. All activities undertaken by both police and community have effects on the dialogue between the two, and thus on the consultative process. Thus, when considering the consultative process as a whole we need to consider the informal initiatives which affect the dialogue between police and community just as much as we need to consider formal consultative groups.

REFERENCES

Brogden, M., Jefferson, T. and Walklate, S. (1988) *Introducing Policework*, London: Unwin Hyman.

Burgess, R. (ed.) (1982) *Field Research: A Sourcebook and Field Manual*, London: Allen & Unwin.

Burney, E. (1985) 'Inside the nick', *New Society* 8 November: 239–40.

Cansdale, D. (1983) 'The development of the Community/Police Consultative Group for Lambeth', unpublished MSc thesis, Cranfield Institute of Technology.

Cochrane, A. (1986) 'Community politics and democracy', in D. Held and C. Pollitt (eds) *New Forms of Democracy*, London: Sage.

Community/Police Consultative Group for Lambeth (1984) *Community/Police Consultative Group for Lambeth: Information Pack*, London: CPCGL.

—— (1987) *Information Pack*, November, London: CPCGL.
Eyles, J. and Smith, D.M. (eds) (1988) *Qualitative Methods in Human Geography*, Cambridge: Polity.
Fyfe, N. (1989a) 'Contesting consultation: the political geography of police–community consultation in London', in D. Evans and D. Herbert (eds) *The Geography of Crime*, London: Routledge.
—— (1989b) 'Policing the recession', in J. Mohan (ed.) *The Political Geography of Contemporary Britain*, London: Macmillan.
Gilroy, P. (1987) *There ain't No Black in the Union Jack*, London: Hutchinson.
Greater London Council (1983) *A New Police Authority for London: A Consultation Paper on Democratic Control of the Police in London*, GLC Police Committee Discussion Paper 1, London: GLC.
—— (1986) *Guide to the Met: The Structure of London's Police Force*, London: Greater London Council.
Hammersley, M. and Atkinson, P. (1981) *Ethnography: Principles in Practice*, London: Tavistock.
Keith, M. (1987) '"Something happened": the problems of explaining the 1980 and 1981 riots in British cities', in P. Jackson (ed.) *Race and Racism: Essays in Social Geography*, London: Allen & Unwin.
London Borough of Lambeth (1981) *Final Report of the Working party into Community/Police Relations in Lambeth*, January, London Borough of Lambeth.
London Strategic Policy Unit (1986) *Police Accountability and a New Strategic Authority for London*, Police Monitoring and Research Group Briefing paper 2, London: LSPU.
Maguire, M. and Vagg, J. (1984) *The 'Watchdog' Role of Boards of Visitors*, London: Home Office.
Morgan, R. (1987a) 'The local determinants of policing policy' in P. Wilmot (ed.) *Policing and the Community*, London: Policy Studies Institute.
—— (1987b) 'Accountability and consultation', *Policing* 3(2): 133–40.
—— (1989) 'Policing by consent: legitimating the doctrine' in R. Morgan and D.J. Smith (eds) *Coming to Terms with Policing: Perspectives on Policy*, London: Routledge.
Morgan, R. and Maggs, C. (1984) *Following Scarman? A Survey of Formal Police/Community Consultation Arrangements in Provincial Police Authorities in England and Wales*, Bath Social Policy Papers: Centre for the Analysis of Social Policy, School of Humanities, University of Bath.
—— (1985a) *Setting the P.A.C.E.: Police Consultation Arrangements in England and Wales*, Centre for the Analysis of Social Policy, School of Humanities, University of Bath.
—— (1985b) 'Called to account? The implications of consultative groups for police accountability', *Policing* 1(2): 87–95.
Scarman, Lord (1981) *The Brixton Disorders, 10–12th April 1981*, Cmnd 8427, London: HMSO.
Walklate, S. (1986a) *The Merseyside Lay Visiting Scheme First Report*, Liverpool: Merseyside Police Authority.
—— (1986b) *The Merseyside Lay Visiting Scheme Second Report*, Liverpool: Merseyside Police Authority.
Weatheritt, M. (1987) 'Community policing now', in P. Wilmott (ed.) *Policing and the Community*, London: Policy Studies Institute.

16

TOWARDS LOCALLY SENSITIVE POLICING?

Politics, participation and power in community/police consultation

Nicholas R. Fyfe

Ever since central government first declared its intention to respond to Lord Scarman's report on the 1981 Brixton riots by implementing his recommendation for local community/police consultation, the resulting police consultative committees (PCCs) have been the focus of political and academic controversy. Political opinion about such consultation quickly polarized between those, such as the then Home Secretary, William Whitelaw, who saw consultation as a mechanism for encouraging policing by consent (*Hansard* 1981: col. 893), and those like the Greater London Council (GLC), who dismissed consultation as 'cosmetic' and falling 'far short of providing a satisfactory system for ensuring the accountability of the police to the community they serve' (GLC 1982: para. 115). Similarly, academic opinion divided between those who believed consultation would act as a form of low-level accountability, making the police more responsive to the expression of community views (Savage 1984), and those who argued that it would be ineffectual because 'at the end of the day [it] leaves policy-making exactly where it was to begin with, firmly in the hands of the police themselves' (Lea and Young 1984: 254; see too Kinsey *et al.* 1986). Although more than four years have now elapsed since PCCs became a statutory requirement in the Metropolitan Police District (MPD) and widely adopted among police forces outside London, studies evaluating consultation remain frustratingly ambiguous about its consequences. Morgan (1989), for example, suggests that PCCs in provincial areas have served to legitimize the tripartite structure for governing the police and to marginalize those groups critical of the police, but acknowledges that such effects may in fact be 'thrown into reverse [by] local factors' (1989: 236).

Against this background, this chapter explores the 'local factors' of community/police consultation. Rather than making generalizations about consultation *across* places and thus leave unexamined those 'local factors' which Morgan identifies as being so important, I aim to look at consultation *in* two contrasting places (the London boroughs of 'Northley' and 'Southam') and thus highlight some of the 'local factors' which make a difference to the structure, process and consequences of community/police consultation.[1] This 'place-centred' (Agnew 1987) perspective on consultation has a number of advantages. First, places provide the contexts for understanding political behaviour, a theme which I shall develop in this chapter in terms of the establishment of community/police consultation in London and in relation to the different participatory strategies adopted by members of local PCCs. Second, a sensitivity to what happens in places means that abstract categories like 'community' and 'power' can be grounded in concrete analysis. The importance of this here is illustrated with respect to the selection of 'the community' for consultation and the 'power' of the police to manipulate the consultative agenda. Third, a comparative place-centred approach highlights the way that similar structures of relations, such as the asymmetries in the power relations between the police and the other participants in consultation, can produce different outcomes in different places (an issue addressed in the conclusion).

PUTTING CONSULTATION IN PLACE

Central government's strategy for implementing the policy of local community/police consultation illustrated a keen sensitivity to the relations between place and politics. In part this was a product of the different political arrangements for police governance which distinguish London from the rest of the country. Outside London, local police authorities were given the responsibility for deciding what if any arrangements should be made for 'obtaining the views of people in that area about matters concerning the policing of the area' (Police and Criminal Evidence Act 1984: s.106(1)). In London, however, where the Home Secretary is the police authority, it would have been impractical for the minister directly to organize local consultation groups and thus responsibility was given to the commissioner of the Metropolitan Police, although he had a statutory duty to follow guidance issued by the Home Secretary and to discuss the form of consultation with the London borough councils. In contrast to the rest of the country, however, consultation was made a statutory requirement in London and was to be based at borough level rather than on an administrative grid structured by police boundaries, the pattern recommended for provincial areas. These details, along with the requirement contained in the Home Secretary's *Guidance on Arrangements for Local Consultation Between the Community and the Police in the Metropolitan Police District* (Home Office 1985, hereon referred to as *Guidance*) that these 'arrangements' take the form of consultative committees comprising an unlimited number of community representatives but with local council representation limited to five, betrayed central government's concern with the distinctive political context of policing in London.

In London, as in many other Labour-controlled metropolitan authority areas, the 1981 urban riots marked a watershed in the local politics of policing (Fyfe 1989b). Councillors questioned their limited, mainly administrative responsibilities for local policing under the Police Act 1964 and many sought to develop a more active role in monitoring and discussing local police activity (see Simey 1982). Nowhere was this commitment to placing policing on the local political agenda more apparent than in London. Here, in contrast to the rest of the country, locally elected representatives have always been denied any statutory responsibility for policing because the Metropolitan Police authority is the Home Secretary. Indeed, successive governments have opposed any reform of police governance in London which would give power to local government because of a belief that policing the nation's capital is too important to be put in the hands of local politicians, a belief reiterated by Lord Scarman in his report on the Brixton disorders: 'I do not believe that Parliament would wish to see ultimate responsibility for the policing of the nation's capital transferred from a senior Minister responsible to it and put in the hands of a local body, however important' (1981: para. 5.68).

When a Labour-controlled Greater London Council came to power in 1981, however, it began a determined campaign to place policing, and in particular police accountability, on London's political agenda. At a regional level this involved a campaign to publicize the way in which the Metropolitan Police were 'out of local democratic control', while at a local level the GLC funded police monitoring groups and encouraged borough councils to set up police committees to campaign locally for greater police accountability to the community. By June 1987 there were nineteen police monitoring groups and fourteen Labour borough councils had established police committees (Fyfe 1989a; 1989c). Against this background, central government's decision to make PCCs a statutory requirement in London and to base them on boroughs can be interpreted partly as an attempt to co-opt into consultation those Labour councillors hostile to the current constitutional arrangements for police governance in London. The GLC's campaign for police reform had thus led to the emergence of a distinctive political geography of policing at borough level, with many Labour-controlled councils demanding greater local government power over the police, while Conservative-controlled boroughs continued to endorse the Home Secretary's responsibility for the Metropolitan Police. These different local political contexts have influenced the participatory strategies of borough councillors in community/police consultation which I explore below.

CONSULTATION IN ACTION: A TALE OF TWO BOROUGHS

'Southam' and 'Northley' are two London boroughs. Southam, a relatively affluent, predominantly middle-class, suburban area with a Conservative-controlled council established a PCC in October 1982. Its structure, decided upon by the council and approved by the local police and the Home Office, allowed for five councillors, three police officers and eight community representatives with meetings to be held in private. This decision to hold private meetings was overturned, however, at the PCC's first meeting by a majority decision of the community membership of the committee. In Northley, a relatively deprived inner city area with a Labour-controlled council, a PCC was not established until February 1985 following lengthy negotiations between central and local government. At issue in these negotiations was the form of community representation. The council argued that as the elected representatives of the people it was best placed to represent the community in consultation through the members of its GLC-style police committee. Central government insisted that this would undermine the independence of the PCC and argued that the PCC should conform

(a) Southam Police Consultative Committee

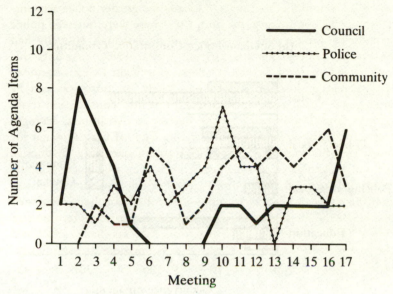

(b) Northley Police Consultative Committee

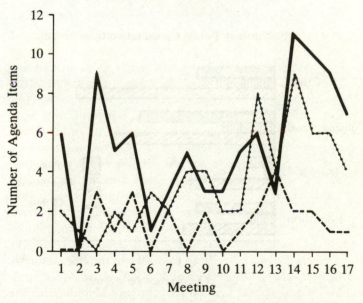

Figure 16.1 Participation in consultation

309

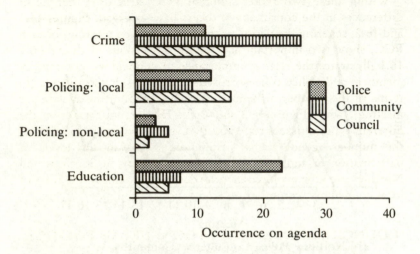

(a) Southam Police Consultative Committee

Occurrence on agenda

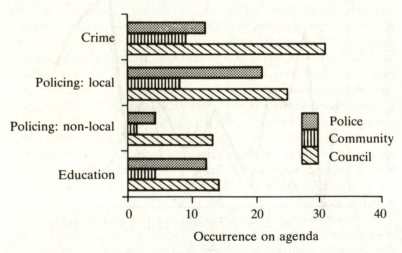

(b) Northley Police Consultative Committee

Occurrence on agenda

Figure 16.2 The content of consultation

to the *Guidance* issued to the Commissioner which allowed for unlimited community membership but restricted council places to five. A compromise was reached, however, whereby Northley council was allowed ten places on the PCC but no restriction was placed on the number of community representatives.

Within these two rather different PCC structures, significant differences in the consultative process have emerged. Figures 16.1 and 16.2, covering the period of the first seventeen meetings of each PCC, show two important and related dimensions of this. Figure 16.1 illustrates the relative importance of councillors, community members and police officers in setting the consultative agenda (in terms of the number of items they each placed on the agenda per meeting), while Figure 16.2 shows the relative importance of the different types of issues raised by PCC members (again in terms of the number agenda items). These differences in the levels of participation in and the content of consultation in Southam and Northley are now explored in more detail by looking at the roles of local councillors, community organizations and police-officers.

COUNCILS AND CONSULTATION: URBAN POLITICS, LOCAL GOVERNMENT AND THE POLICE

It is evident from a comparison of Figures 16.1a and 16.1b that there were important differences between Southam and Northley in the level of council participation in consultation. Although in both boroughs councillors began by playing the dominant role in setting the consultative agenda, only in Northley was this sustained over the study period whereas in Southam there was a period of about a year when there were very few contributions by councillors. There were also important differences between the two boroughs in terms of what was discussed (see Figures 16.2a and 16.2b). In Southam council interests focused around local policing policy whereas in Northley it was clear that the Labour councillors had a much wider set of concerns, ranging from crime (its incidence and prevention), both local and non-local policing policy, and education (using the consultative arena to educate the community about different policing issues).

In understanding these different levels and forms of participation it would be tempting to invoke a simple party political explanation: the relatively passive involvement of Tory councillors reflecting a belief that there should be minimal party political influence in

311

policing; while the active role of Labour councillors being accounted for in terms of their commitment to developing the accountability of the police to local government. Such an explanation is flawed, however. Closer examination of the local history of council participation in Southam reveals that the passivity of Tory councillors reflected the fact that they boycotted consultation having claimed that it was ineffective in influencing police thinking; while reference to the geography of Labour council participation in London reveals that Northley council was exceptional in terms of taking an active role in consultation because many other Labour councils believed that participation would undermine more radical attempts at police reform. Against this background any explanation of council participation in consultation must be more sensitive to 'local factors', that is to the interrelationships between place and politics.

Community/police consultation in Southam posed a political dilemma for the ruling local Tory councillors. On the one hand, they supported central government's opposition to any change in police governance in London which would transfer power to local politicians, one councillor commenting about police accountability that 'I think that London's got it about right and the rest of the country has got it wrong' (interview). On the other hand, if policing at borough level was to be discussed then the Tory councillors believed that they as the local ruling party and the elected representatives should occupy a privileged position in community/police consultation. The tension between these two perspectives became manifest soon after consultation began in the borough. Although many Tory councillors went into consultation believing that it was 'a waste of police time' and that the police should simply 'get on with their job', following a presentation by the police of problems in the borough at the first meeting, the Tory councillors became concerned by a conflict between what they perceived local police priorities should be – street violence and criminal damage – and the police statement that domestic burglary was their main priority. Despite repeated requests at subsequent meetings by Conservative councillors to the police to allocate more of their resources to dealing with the council's priorities, the police responded by emphasizing that with limited resources they were unable to give these offences priority. Furthermore, at the first meeting of the Southam PCC the community membership had overturned the council's decision that the PCC hold its meetings in private and instead made the meetings public but reserved part of the agenda for confidential items. At a subsequent meeting this meant that Tory councillors in the public gallery were asked to leave for a confidential item. This prompted a demand by the council that the constitution of the PCC be changed to allow more councillors to be members, a demand refused by the community membership of the PCC. This in turn triggered a walk out by all the Tory councillors, who were angry that their authority as elected representatives was being undermined both by decisions taken by unelected community representatives and by the police being unresponsive to their demands. Tory councillors returned only after they had negotiated an agreement with the PCC that allowed an increase in council places.

It is in Northley, however, that councillors consistently played the dominant role in setting the consultative agenda despite community representatives having the majority of places on the PCC. The reasons for this were closely associated with the explicit commitment made by Northley's Labour-controlled council back in 1982 to take an active role in local crime and policing issues by following the lead of the GLC and establishing a police committee. Consisting of fourteen councillors supported by a research unit staffed by council officers with experience of criminological and police research, the police committee initially identified its priority as campaigning for greater police accountability to the local community. This aim corresponded with the position of the GLC and other London Labour council police committees but unlike them Northley councillors decided early on that the campaign for police accountability was a necessary but not a sufficient cause to pursue:

> if local authorities wish to play a greater role in policing, they will have to not only campaign for greater police account-ability but will also have to introduce a policing variable into other policy areas. ... Local authority policy in areas such as housing and social services has a direct bearing on policing, for example the installation of flood lights on council estates not only reduces peoples' fear of crime but also reduces the commission of offences like burglary and vandalism on that estate.
>
> (Report of Northley Council Police Committee Support Unit, July 1983)

The development of a council co-ordinated crime prevention strategy and the campaign for police accountability thus became important issues for the police committee and these priorities fed into council involvement in the PCC.

At the early meetings councillors regularly used the consultative group as a public arena in which to highlight the lack of account-ability of the police to the local community by raising non-local policing issues which had important implications for Northley and by educating the community about aspects of police organization (see Figure 16.2b). In raising the issue of the policing of the 1984–5 miners' dispute, for example, councillors condemned the way it involved taking police-officers away from the local area for use on picket line duty in the coalfields 'without regard for the policing needs of those areas which are understaffed as a result of the strike'; while a presentation by council officers of the police committee illustrated the powerlessness of the community in relation to decisions about police use of riot control weapons. Over time, however, there was an important shift in the participatory strategy of the Labour councillors in the Northley PCC, away from conflict with the police about local accountability and towards co-operation with the police in local crime prevention initiatives. This became a major concern of councillors (see Figure 16.2a) and resulted in a number of concrete projects such as two multi-agency crime prevention schemes on council housing estates and the production of a leaflet with the police about the law on squatting.

Although this shift in participatory strategy connected with the council police committee's interest in crime prevention, the timing of it reflected the re-election of a Conservative government at the 1987 General Election, which meant that the possibility of an elected police authority for London would not be on the national political agenda for at least the next five years. The implications of these national political developments for local political relations was spelt out by a Labour councillor on the police committee:

> You've got to be realistic. We've convinced the Labour Party of the need for an elected police authority in London but until they're in power we haven't got a hope of that happening. So should we bang our heads against a brick wall until then? We've got to show the police that we are capable of working with them.
>
> (interview)

Furthermore Northley Labour councillors praised the PCC, argu-ing that it had encouraged a degree of what they called 'moral accountability' of the police to the community ('if so and so is shouting at them to do something about women being attacked then there isn't another argument, they've got to say they will do something': interview); and that it played a useful role in building up a knowledge of police procedures which would be important if local government was to play a more active role in police affairs under a future Labour government.

'THE COMMUNITY' AND CONSULTATION

In its *Guidance* on community/police consultation, central government claimed that consultation would allow 'general policing policies to be adapted to meet identified needs in the light of the expressed wishes of the local community' (Home Office 1985: para. 3). Underlying this claim, however, is an essentially pluralist conception of 'the community' as some kind of pre-existing, spatially defined entity which can be 'represented' through the participation of a number of willing individuals. As I intend to show in this section, the grounding of central government's concept of 'the community' in particular places reveals that community involvement in PCCs has been very much at odds with central government's pluralist perspective.

The belief that consultation would 'reach into the local community [and] draw in the people who live there to reflect the views of their own neighbourhood' (Home Office 1985: para. 3) contrasts with the way in which London borough councils have played key 'gatekeeping' roles in terms of deciding the composition of 'the community' present at PCCs. In Southam, for example, the council decided unilaterally on the initial composition of the PCC drawing members principally from the local chambers of commerce and the council for voluntary services. This choice of community organizations was far from arbitrary and in fact stemmed from a particular definition of 'the community' in terms of those 'approved' by the council. As one Tory councillor commented:

> We invited all the groups we knew were performing a function within the community and we drew the line at other organisations which we regarded as overtly political such as trades unions or what I would call pressure groups for just one thing, such as residents associations.
>
> (interview)

315

Table 16.1 Community membership of police consultative committees 1986–7
(a) Southam Police Consultative Committee

	Gender		Race		Age	
	Male	Female	White	Ethnic	Under 45	45+
Voluntary service (1)	x					x
(2)		x	x			x
Youth groups (1)	x		x		x	
(2)	x	x	x		x	
(3)			x	x(A)	x	
Ethnic groups	x		x		x	
Church	x		x		x	
Trade/commerce (1)	x		x			x
(2)	x		x			x
Old people	x		x			x
Neighbourhood watch		x	x		x	
Victim support		x	x			x
Residents' Assocs. (1)	x		x			x
(2)		x	x			x
(3)	x		x			x
(4)	x		x			x
(5)	x		x			x
Total (number)	12	5	16	1	6	11
(percentage)	71	29	94	6	35	65

(b) Northley Police Consultative Committee

	Gender		Race		Age	
	Male	Female	White	Ethnic	Under 45	45+
Tenants' assocs. (1)		x	x		x	
(2)	x		x			x
(3)	x		x		x	
(4)	x		x			x
(5)		x	x			x
Ethnic groups (1)	x			x(A/C)	x	
(2)	x			x(A/C)	x	
(3)	x			x(A)	x	
(4)	x			x(A)	x	
(5)	x			x(T)		x
Victim support		x	x		x	
Trade/commerce	x		x			x
Church	x		x			x
Neighbourhood watch (1)	x		x		x	
(2)	x		x			x
Total: number	12	3	10	5	8	7
percentage	80	20	67	33	53	47

Key: A-Asian A/C-Afro-Carribbean T-Turkish
Source: Fyfe 1989c

In Northley, too, Labour councillors initially selected a 'community' of which they 'approved', giving the majority of places to ethnic minority groups (all of which were council-funded organizations) and tenants' associations (for whom the council is their landlord). Once the PCCs were established, the police too began to play an important 'gatekeeping' role by encouraging groups which they believed to be supportive of the police to seek membership. In Northley, for example, senior police-officers actively sought the involvement of neighbourhood watch and pensioner organizations.

The selection of individual representatives from within community organizations invariably involved a process of the (self) appointment of an existing office-holder (the secretary or chair, for example) rather than their election to a position on the consultative group (Fyfe 1989c) and, as the profiles of these community representatives shown in Tables 16.1a and 16.1b illustrate, there were important biases in terms of the age (most were over 45), gender (most were men) and race (most were white) composition of the community membership of the PCCs in Southam and Northley. Furthermore, the biases of age, gender and race were reinforced by levels of attendance at and participation in consultation. In Southam, for example, residents' association and voluntary service representatives each placed twelve issues on the consultative group's agenda over the study period compared to four items by the ethnic minorities representative; while in Northley the five tenants' association members raised nine issues compared to three by the five ethnic minority group representatives. The importance of these biases can be gauged by the fact that as a number of crime surveys have now shown these groups are generally the most supportive of the police (see Jones et al. 1986).

To take this analysis of community involvement in consultation further, the participatory strategies available to community representatives can be understood to vary according to whether representatives adopted an active or passive role, and whether they were supportive or critical of policing. This yields four 'ideal' types of participatory strategy which can be used to characterize the involvement of different community organizations. A passive but supportive role was characteristic of some residents' representatives in Southam and Asian representatives in Northley. Both these groups described their relations with the police as being consensus- rather than conflict-based and had good informal links with local police-officers. But passivity was also associated with a critical attitude towards policing and this characterized the non-participation of Afro-Caribbean groups in Northley. Their reluctance to participate was a product of a distinctive local history of relations with the police: 'the kids round here have got the scars of years of police harassment, you know the physical scars of a criminal record at an early age and of where the handcuffs were put on' (interview). Such experiences led one Afro-Caribbean to dismiss the PCC initiative as simply a cosmetic exercise, of only superficial significance when compared with the deep-seated conflict which had emerged between the police and sections of the community. He explained that relations with the police were not a *community* problem but a *police* problem which required changes in police behaviour, not community involvement in what they perceived as powerless consultative groups. Another Afro-Caribbean representative in Northley feared being manipulated by the police given the way in which membership of the PCC conferred the status of a 'community representative' on the participant:

> If I start going to the group the police would use me when they had any trouble with the black community; they would come to me as a 'community leader' but I don't represent anybody, I'm just in charge of this club.

> (interview)

An active and supportive role was characteristic of voluntary service and neighbourhood watch representatives in Southam who encouraged the police to 'educate' the community about issues such as the role of the special constabulary and home beat policing. An active but critical role was characteristic of many of the tenants' and residents' associations in Northley and Southam, who used their PCCs as a forum to which they would bring complaints about the incidence of crime (see Figures 16.2a and 16.2b), usually of minor offences such as parking or nuisance behaviour by youths. This often led to heated exchanges because of the way in which the police were perceived by the community representatives to respond to their demands by hiding behind claims of lack of resources: 'You get nothing from the police. I haven't seen a home beat around here for years but when I told them at the consultative group they just said "There's a limit to what they can do"' (interview). An important implication of this concerned the way in which community/ police consultation had clearly raised the expectation among community participants that 'something' was being done to make the police more responsive to local demands but when these demands appeared to be ignored, community participants became more critical of the police than before.

This issue of police responsiveness to community demands is clearly of pivotal importance to community/police consultation. According to central government's *Guidance*, consultation would allow 'general policing policies to be adapted to meet identified needs in the light of the expressed wishes of the local community' (Home Office 1985: para. 3). However, the pluralist assumption underlying this claim, that participation is power, is fraught with problems. Clearly, policing decisions are in fact ultimately determined by the law rather than community 'wishes' given that the constitutional basis of policing in this country is founded on the principle of constabulary independence, 'the unique obligation on the police to uphold the law under powers which are granted by law, to which the office is accountable' (Jefferson and Grimshaw 1984: 15). The primary importance of this principle leads Jefferson and Grimshaw to dismiss community/police consultation as 'incoherent':

> Any system of policing focused simply on the principle of law enforcement – the unconditional requirement to enforce the law – cannot readily admit other principles – such as a requirement to operate only with community consent – without risking incoherence.
>
> (Jefferson and Grimshaw 1984: 102)

Within the structural limits set by the principle of constabulary independence, however, there remain important issues at stake which influence the quantitative and qualitative aspects of policing delivered to the community. At present, however, decisions about, for example, the number and distribution of officers, or the importance given to public order policing as opposed to criminal investigations are still specifically excluded from the agenda of community/police consultation by central government. Only within these formal (legal) constraints set out above can community participants in consultation potentially exercise any influence over the local discretionary decision-making of the senior police officers who attend consultative group meetings. In the next section I shall consider the power exercised by these police officers in community/police consultation in the context of their responsiveness to community 'needs' and 'wishes'.

THE POLICE AND CONSULTATION: THE POWER AND AUTONOMY OF URBAN MANAGERS

The image developed by the senior local police-officers (chief superintendents and superintendents) who routinely attend PCCs has been that of the urban manager, a strategically placed allocator and controller of scarce and desired urban resources (Saunders 1981: 118–19; Pahl 1975). Demands made by community participants and councillors for the police to deal with, for example, the enforcement of a heavy lorry ban, parking problems around the local football ground or the fear of crime in a pedestrian underpass, have required these senior officers to take decisions about whether or not to allocate police personnel to deal with these locally perceived problems.

In taking such decisions, however, senior officers have a degree of autonomy (in relation to superior ranks) and a degree of discretion (in relation to the law) which prohibits any simple generalization about their response to the wishes of PCC members. This can be illustrated partly by an inter-divisional comparison of their actions and decisions in response to similar council and community wishes in the two boroughs. The refusal of the chief superintendent in Southam, for example, to reveal details to the PCC of the number of officers used outside the borough for policing industrial disputes (because he believed it was an 'operational matter') contrasted with the decision of the chief superintendent in Northley to make these figures available to the PCC. Furthermore, the refusal of the police in Southam to respond to the wishes of councillors and give priority to problems of vandalism and street robbery contrasts with the success of Northley councillors in persuading the police to change their local priorities (as set out in the police divisional annual reports) to correspond with those identified in a crime survey (organized and funded by the council) such as sexual assaults, domestic burglary, street robbery and heroin dealing.

These local differences in police responsiveness were largely a product of the discretionary behaviour of senior officers and in particular their attitude towards 'doing business' with local councils. In Southam the local chief superintendent believed that it was illegitimate for the police to have anything to do with 'party political initiatives'; whereas in Northley the senior officers actively sought to harness the council's resources to local crime and policing projects such as the crime prevention projects on two housing estates (see pp. 313–14).

However, these differences in approach should not be allowed to conceal the underlying asymmetrical power relations which structure consultation between *any* community and the police. At one level this is manifest clearly in the legal power of the police (enshrined in the doctrine of constabulary independence discussed above) to determine the outcome in situations where there was an open conflict of interests about, for example, local priorities in the allocation of police resources. This represents what Lukes (1974) refers to as the one-dimensional exercise of power. But relations between the police and the community in PCCs also disclosed what Lukes refers to as two- and three-dimensional power relations. The exercise of two-dimensional power by the police was evident in their control over the agenda for discussion and thus their ability to prevent decisions from being taken on potential issues over which there was a conflict of interests. In part this is illustrated by the way local police-officers specifically prohibited discussion of some issues. The Southam divisional chief superintendent, for example, specifically refused to discuss precise 'manpower levels' in the borough despite repeated requests by members of the PCC that they be kept informed of the deployment of officers. Control of the consultative agenda by the police also occurred in more subtle forms, however, by manipulating issues such that comments critical of local police procedures were occasionally interpreted as a formal complaint and thus not for discussion in the consultative arena.

The exercise of police power in Lukes's three-dimensional sense was evident in the way the perceived interests of some community participants appeared to be gradually altered so as to coincide with those of the police. This was closely associated with an education process by which community participants came to accept police claims about limited resources and so tempered the demands they made of the police. This was most apparent in Southam where, as Figure 16.2a illustrated, the police adopted a proactive role in educating the PCC. The implications of this were illustrated when a community participant was asked about his suggestions for the local divisional policing plans:

> It's very difficult to suggest anything because they are so limited as to the number of people they can use. If you start putting men on specific areas then you are going to lose out in other areas. Yes, you would like them to take action in some areas but you've got to realise they just haven't got the manpower.
>
> (interview)

Indeed, Morgan (1987: 35) identified a similar process occurring in provincial consultative groups: 'Given [police] accounts of serious crime it begins to seem reasonable to audiences that the police should give low priority to such matters as youths cycling on footpaths.'

This power of the senior divisional police-officers to control the consultative process clearly tends to reinforce the image of these officers as relatively autonomous urban managers. Nevertheless, consultation has also provided evidence of how this image and the responsiveness of these officers to community demands can be seriously compromised by the decisions and actions of officers both *above* and *below* them in the police hierarchy. Indeed, it is precisely those problems with the managerialist thesis which Pahl (1975) identified in his later work: those of the *autonomy* of urban managers (given the constraints imposed by the wider structure) and the *identity* of urban managers (given that it is lower-level employees who actually deliver the service), which have intruded into consultation to expose the vulnerability of chief superintendents and superintendents as urban managers.

The problem of the autonomy of senior local officers in relation to the higher echelons of the force was most clearly evident in conflicts over decisions at force level to take police personnel away from a division for use in the policing of, for example, industrial disputes. In such situations the local community and the local police found themselves powerless to prevent this. Indeed, one divisional chief superintendent even pleaded with the consultative group to make representations to headquarters on his behalf to stop them taking officers from the division: 'Area is thinking of taking manpower away from the division – if this happens we will come back to you to put pressure on Area'.

The power of senior local officers as managers of a division has also been shown to be acutely vulnerable to the actions of lower-ranking officers. A regular complaint made to senior officers by community and council participants in consultation was that 'What you tell us here just doesn't seem to be heard at street level, things just aren't filtering down to those police officers.' As one community representative observed in relation to the police response to racial attacks:

My view of the police response is that the police at a higher level, those at the Group who are just a little unit, they are very sensitive and very helpful. *But at the groundfloor level, things are not happening.* When a policeman comes to the door to investigate an incident, he or she is not interested in the racial factor, they don't care.

(interview, my emphasis)

Indeed, this perception of a gulf between what is said at consultative groups and what happens on the streets was corroborated by discussions with police sergeants and constables. Although senior officers in both Southam and Northley claimed that the substance of consultative group meetings was communicated to other officers, interviews with lower-ranking officers revealed that very few could recall issues which had been discussed. Indeed, although all knew of the existence of PCCs, overwhelmingly the attitude was that these 'don't have anything to do with my job', justifying this in terms of other, more immediate demands made on police time by the public and, more particularly, the discretion patrol officers have in carrying out their duties (Fyfe 1989c).

The evidence presented in the latter part of this section indicates that in addition to legal constraints there are some important organizational and occupational constraints on police responsiveness to community 'needs' and 'wishes' which call into question not only the autonomy of senior officers at PCC meetings but also their *identity* as urban managers. Rather than being 'independent' variables allocating scarce police resources they appear more as 'intervening variables' between policy decisions taken at area or force level and the discretionary actions of officers on the street.

CONCLUSION: TOWARDS LOCALLY SENSITIVE POLICING?

This comparative study has attempted to illustrate some of the 'local factors' of relevance to the conduct of community/police consultation. Although similar asymmetrical power relations clearly existed in both Southam and Northley between the police on the one hand and councillors and community organizations on the other, the local consequences of consultation have differed. In Southam there was little evidence of the police being responsive to council or community demands and only one joint police–community initiative emerged over the study period: the production of guidelines on the sale of alcohol to young people. Meetings of the Southam PCC were characterized by discussions of the incidence of relatively minor offences and presentations by police-officers on matters of local interest. By contrast, in Northley not only have the police been responsive to some council and community demands but also major police–council initiatives have been developed in the consultative arena, such as the multi-agency crime prevention projects on two council estates. These initiatives were a product of the way in which the council's police committee harnessed its resources and expertise to the consultative group and the willingness of local senior police officers to participate in these projects.

Nevertheless this chapter has also attempted to illustrate the limitations of community/police consultation for ensuring that the police respond sensitively to community wishes. These limitations are in part legal (given the doctrine of constabulary independence), organizational (given that many of the key decisions which affect local policing are taken by high-ranking personnel absent from the consultative arena), occupational (given the capacity of 'street cops' to subvert decisions taken by 'management cops') and political (given that the community representatives are largely unrepresentative with no local democratic mandate). Against this background community/police consultation is still a long way from fulfilling central government's hope that it would allow 'general policing policies to be adapted to meet identified needs in the light of the expressed wishes of the local community' (Home Office 1985: para. 3).

Acknowledgements

I would like to thank Kevin Heal and John Lowman for comments on an earlier draft of this chapter. The research on which this chapter is based was funded by the Economic and Social Research Council.

NOTE

1 This chapter is based on research carried out in 1986–7 and draws upon an analysis of the minutes of the two consultative committees, in-depth interviews with committee members and observations at committee meetings. It is part of a larger project described in Fyfe 1989c.

REFERENCES

Agnew, J. (1987) *Place and Politics*, London: Unwin Hyman.

Fyfe, N.R. (1989a) 'Contesting consultation: the political geography of community/police consultation in London', in D.J. Evans and D.T. Herbert (eds) *The Geography of Crime*, London: Routledge.

—— (1989b) 'Policing the recession', in J. Mohan (ed.) *A Political Geography of Contemporary Britain*, London: Macmillan.

—— (1989c) 'Community/police consultation in London and the political geography of policing', unpublished PhD thesis, University of Cambridge.

Greater London Council (1982) *The Policing Aspects of Lord Scarman's Report on the Brixton Disorders*, London: GLC.

Home Office (1985) *Guidance on Arrangements for Local Consultation Between the Community and the Police in the Metropolitan Police District*, London: Home Office.

Jefferson, T. and Grimshaw, R. (1984) *Controlling the Constable: Police Accountability in England and Wales*, London: Muller/Cobden Trust.

Jones, T., MacLean, B. and Young, J. (1986) *The Islington Crime Survey: Crime, Victimization and Policing in Inner City London*, Aldershot: Gower.

Kinsey, R., Lea, J. and Young, J. (1986) *Losing the Fight Against Crime*, Oxford: Basil Blackwell.

Lea, J. and Young, J. (1984) *What is to be Done about Law and Order?*, Harmondsworth: Penguin.

Lukes, S. (1974) *Power: A Radical View*, London: Macmillan.

Morgan, R. (1987) 'The local determinants of policing policy', in P. Wilmott (ed.) *Policing the Community*, London: Policy Studies Institute.

—— (1989) '"Policing by consent": legitimating the doctrine', in R. Morgan and D.J. Smith (eds) *Coming to Terms with Policing: Perspectives on Policy*, London: Routledge.

Pahl, R. (1975) *Whose City?* 2nd edn, Harmondsworth: Penguin.

Saunders, P. (1981) *Social Theory and the Urban Question*, London: Hutchinson.

Savage, S. (1984) 'Political control or community liaison?', *Political Quarterly* (1).

Scarman, Lord (1981) *The Brixton Disorders 10–12 April 1981*, Cmnd 8427, London: HMSO.

Simey, M. (1982) 'Police authorities and accountability: the Merseyside experience', in D. Cowell, T. Jones and J. Young (eds) *Policing the Riots*, London: Junction Books.

INDEX